Motion Unification

The Narrative

Maxwell Hart

authorHOUSE®

AuthorHouse™
1663 Liberty Drive
Bloomington, IN 47403
www.authorhouse.com
Phone: 1 (800) 839-8640

Published by AuthorHouse 05/12/2016

ISBN: 978-1-5049-6687-0 (sc)
ISBN: 978-1-5049-6688-7 (hc)
ISBN: 978-1-5049-6686-3 (e)

Library of Congress Control Number: 2015920785

Print information available on the last page.

This book is printed on acid-free paper.

Contents

Excerpt

"Physics has come a long way in the 300^+ years since Newton declared gravity to be a natural force. The failure to find the source of this natural force or the nature of this force of nature is not for lack of trying. The search goes on even today. Does the Higgs boson ring any bells? What about the search for gravitation waves? If the source of gravity revealed in this document is certified, then this one discovery will simplify and motivate much new searching and researching by many new searchers and researchers."

1. Introduction

Caution

Please do not decide whether this book is beyond your understanding until you have finished reading the next six to ten pages. It will get easier as you get used to it.

This book was not designed for just the professionals. The language is mostly on the high school level and, with a web browser and a little patience, many junior-high school students should be able to understand the basic principles. Occasionally, throughout the book, you will find euphemisms (synonyms) in parentheses that might be of assistance to some younger readers.

Most of the technical terms and the new words that are introduced in this tome (scholarly book) are also contained in the glossary at the end of the book. You might want to make a copy of those pages to reduce the number of times you will find it necessary to turn to the back. The use of a tablet or smart phone might simplify the copying process.

Getting Started

The excitement of the many new discoveries to be revealed in MOTION UNIFICATION does not really start until Chapter 2. However, most readers will have some difficulty navigating the details of some of these revelations without at least tolerable understanding of many of the definitions that are presented in this ***Introduction***.

You will soon discover that there are many new terms defined for the first time in this book that cannot be looked up anywhere else (yet). You will also discover that the concepts presented in this book are not completely independent.

Many concepts carry over to (are assumed in) succeeding concepts. However, anyone who becomes bored with the definitions of the ***Introduction*** might be justified in taking a precipitous peek at Chapter 2 before returning to this ***Introduction*** whenever vocabulary bolstering seems prudent.

Any adult reader only wishing to learn how gravity works might be able to understand enough about chapters 1, 2, and 7 to catch the essence of it. If that goes well, then chapter 9 will fill in some of the details and provide some of the consequences.

Natural Behavior

Before Sir Isaac Newton "discovered" gravity (back in the late 17th century), the falling of objects had been considered well understood. Falling was the natural behavior of objects that were dropped or otherwise unsupported. Newton's fortuitous discovery of gravity illustrates a fundamental truth that is not fully appreciated even today.

Any "natural behavior" that is "well understood" should be considered with suspicion until "how it works" is also well understood.

Lest we be too judgmental of a supposed early lack of criticality, we might now ask ourselves, for example, "How does momentum work?" Is this just natural behavior that is well understood? Isn't anyone curious about ***how*** it works? Will future generations of physicists criticize our lack of criticality?

Some might say that physics has ***always*** been about how things work physically. Before we can talk about "how it works," we must first establish our intended meaning of these three words. In physics, we tend to use formulas to simplify our explanations of how various components of a system are related. When all of the formulas of a particular system are understood, we can say that we know how the system works because we can make accurate predictions.

However, there are levels of knowledge. We will condense this down to three levels. The first level requires observations to explain what the system does. The second level requires mathematical formulas to model the variables. The third level requires a theory that physically explains, as simply as possible, the principles of how the system works. Equations and formulas will be used only when system relations must be established before the system can be analyzed.

In the case at hand, we do have formulas for momentum that allow us to make accurate predictions. However, in this case there is nothing to explain. Momentum is just natural behavior that is well understood! Right ...? Let's exercise some critical thinking.

Example of "How It Works"

Level 1: It is possible for a child to observe an old but still functioning mechanical alarm clock for a few days to discover what it does.

Level 2: By studying the clock and experimenting with all of its knobs and dials (or reading the directions taped to the back if all else fails), it is possible for a young mathematician to model the relationships between the input and the output. Models can often be revised and restated in many ways until it is generally agreed that it has been expressed as simply as possible.

Level 3: However, if clock making technology were ever lost, it might require a theoretical physicist (or in this case an experienced mechanic or engineer) to take the clock apart. In a process called reverse engineering, it might be possible to gain a real <u>understanding</u> (theory … not conjecture) of the internal operation of this particular clock that physically confirms or alters the mathematical models and discerns the simplest theory of how it is supposed to work.

There are at least four terms that, depending on the source of the term, imply a variety of credence. In this book, these terms will be understood as follows.

Conjecture – an opinion or guess in the form of an argument.

Premise – a conjecture, hypothesis, or assertion generally given to establish an exact statement of some argument to be studied with a goal of possibly deriving a valid proof of that argument.

Theory – an argument or inference that has passed every test and is generally accepted as true or valid. (An argument is "valid" if the premises support the conclusion. In other words, in a valid argument, premises, if shown to be true, would guarantee a true conclusion.)

Narrative – the theory with the highest Occam and "strongest heart" (to be defined soon) that has been fortified by an understanding of how it works. This will generally be the theory that is most acceptable as an understanding of how it works.

There will be many examples of these technical terms used throughout this book.

The Narrative Principle

While occasional fluctuations may be difficult to fathom, standard behavior occurs for a physical reason that can be expressed as a "narrative."

Reasoning

Mathematics is a critical tool for analyzing data and synthesizing formulas that can be used for predicting future behavior. Many mathematical and scientific discoveries start with one or more premises and follow the patterns or implications logically, using previously established definitions, theorems, or physical laws to prove conclusions directly. This type of reasoning is called "deductive" reasoning.

"Inductive" reasoning, however, uses **strong** evidence such as numerous independent examples and known behavior to support its premise(s). Following the implications of such evidence, a **narrative** can sometimes be developed that supports the conclusion(s) of the argument.

Any **narrative** that leads to or supports another independently valid **narrative** tends to establish both as significant. The greater the supported connections, the greater the significance of all related concepts. *You will find this "**MOTION UNIFICATION**" filled with mutually supporting* narratives.

The greatest danger with inductive reasoning is the possibility of applying a chain of evidence based furtively on one or more of the argument's premises or even on its conclusion to support an assumed premise of the argument. This is called circular reasoning. As you will discover in this book, critical examples of circular reasoning still exist in our modern understanding of physics.

Circular reasoning is sometimes tolerated because it has the same feel as reinforcing narratives. There is a huge difference, however. Circular reasoning actually **thwarts** reinforcement.

When two competing concepts are developed inductively, the simpler concept with the fewest and *least restrictive* requirements is generally considered loftier due partially to its greater simplicity, but primarily to its wider (*less restricted*) field of relevance and consequently superior reinforcing evidence.

Regal Science

Any thinking that is given a pass on ***understanding*** the **narrative** of how it works will be called "regal science" because of its treatment as fiat. The circumstances may be unusual, but the consequences of regal thinking are illustrated by the fable, **THE EMPEROR'S NEW CLOTHES** by Hans Christian Andersen.

One of the best-known physics examples of regal science is Aristotle's geocentric solar system from the 4th century B.C. The strategy was simple enough and the original model was as simple as possible. It was, in fact simpler than today's models. However, the heart of the **narrative** was weak. As the data became more precise, the necessary adjustments to the model became more complicated.

Four hundred years ago, Johannes Kepler announced his empirical laws (discovered experimentally) that accurately described motions of the planets. Three hundred years ago, Sir Isaac Newton improved Kepler's laws and provided many of the mathematical formulas that allow predictions concerning gravitational orbits. Last century, Albert Einstein hypothesized that gravity travels between massive objects in gravitation "waves."

Yet, from the field of physics, the mystery of gravitation, where it comes from, and how it works has eluded us until now. Even hypothesizing a "God particle" has not helped. Currently, the realm of regal science includes (but is not limited to) NEWTON'S THREE LAWS OF MOTION, MOMENTUM, ANGULAR MOMENTUM, PRECESSION, COSMIC FLATNESS, COSMIC ACCELERATION, MASS, GRAVITATION, and CONSTANT C.

Establishing these **narratives** will allow several long-standing physics mysteries to be finally put to rest in this book. The physical validation of **narratives** will include *THE HEISENBERG UNCERTAINTY PRINCIPLE, FABRIC OF SPACE, DARK ENERGY,* DARK MATTER, *VACUUM ENERGY, ISOTHERMAL ENERGY, THE EQUIVALENCE PRINCIPAL,* two new *EQUIVALENCE PRINCIPLES, GRAVITATION WAVES,* (similar) *PRECESSION WAVES* and *NIELUCION WAVES*, and many others.

A new concept called "orbitar precession," or "nielucion" will also be revealed for the first time even as it hides in plain sight. *(Note that the spelling and pronunciation of "orbitar" is similar to that of "linear" and "angular." Nielucion will be revealed and discussed in* ***"Gravitation Concept.")***

In the inductive presentations of this ***MOTION UNIFICATION***, there will be numerous instances where multiple options are possible. The development of **narratives** will be used to explain how particular options work.

Systems with physical **narratives** will generally be elevated above the ones with only mathematical formulas for justification. As stated earlier, the option with the fewest and simplest assumptions allows the broadest application and greatest reinforcement. This option will be called "the simplest Occam." Hypotheses are always assumed true. Inductive arguments, as this presentation must be, are seldom supported by deductive proof but by the strength and productivity of their inductive conclusions and by the simplicity and paucity (divined by Occam's razor) of their assumptions.

Such arguments must be revised or abandoned if contradictions occur or predicted consequences do not. "Occam's law of parsimony" is named in honor of Franciscan friar William of Ockham.

MU will be using Occam's razor to develop a deeper, more insightful version of physics that might be called "**narrative** physics." This realm of physics will seek to understand the physical reality beyond the mathematical models of many well-known phenomenon.

Absolute Motion

Having found the simplest system that works in a particular setting does not generally entitle us to say that this must be how nature behaves in all settings and that all contrarian theory is incorrect.

For example, since the geocentric (earth centered) model of the solar system has been replaced with the heliocentric (sun centered) model, it is sometimes suggested that we no longer say (or even think) that the sun rises in the east and sets in the west. The "correct view" is that the earth rotates counter-clockwise making it appear from the surface of the earth that the sun rises and sets.

Nevertheless, we should not underestimate the importance of appearances. In this regard, many social scientists and PhDs insist, "perception is reality."

Wisdom leads us to the discovery that we must often choose between two or more different ***versions*** of the "reality" of our perception. Perception without understanding leads to "certainty" which leads to dogma. Perception with understanding leads to the confidence that gives us the courage to question our perceptions, reevaluate our "reality," and broaden our horizons.

The Simplest Occam Depends On the Point of View

When studying the solar system, the heliocentric model is usually simplest. When discussing the mechanics of lunar or other earth satellite motion, the geocentric model is usually simplest. When considering travel through the universe, both of these models could become inadequate and some galactic model might be simplest.

The Strongest Heart

There is a problem with making the simplest Occam our ultimate goal for the improvement of theory. Sometimes, in order to work, seemingly simple concepts must be construed to form logical systems that are more complicated than necessary. Constant c and relative motion seem simple. Even space warp seems simple at first.

Consider the circular orbits of Aristotle's geocentric solar system. The elliptical orbits of today's models are more complicated, but the circular orbits combined with the necessary (perspective) adjustments make Aristotle's model far more complicated than necessary.

It is important that the "core" of a logical system be simple and that all parts are expressible in terms of that context. It is even more important that the "heart" of the system be fathomable. The heart of a logical system can be judged by the simplicity of its narratives.

At the end of the day, ***the best physical theories will be those with the "strongest heart."*** **MU** will always attempt to use narratives to differentiate those systems with strongest heart.

Principium Axioma
(The Axiomatic Principle)

Every branch of mathematics, science, and most other logical systems are "axiomatic systems." To emphasize our definition of axiomatic systems (below), we will call an axiomatic system an "asys" for short. This might be pronounced "aces". The plural would be spelled "asies," and pronounced, "a-sea´s."

To study any asys, the assumptions made determine the asys being studied as well as the point of view of the study. These assumptions are sometimes called postulates or axioms. Any distinction between these two words is no longer considered important.

Each of the three solar system models mentioned above can produce a valid asys within the framework of the corresponding system. No asies and no theorems proven in any asys have any validity when considered within any alternative asys unless reproved within that alternative setting. In **MU**, this ***crucially important*** property of all asies will be christened "principium axioma" to distinguish references to it from references to other general properties of asies and references to specific asies and to specific properties of specific asies.

The Motron Hypothesis (MH)

Photons are considered by particle physicists to be elementary particles called gauge bosons* that seem to have these three qualities.

- They have no charge and no mass.
- Each is a quantum of light.
- They are the carriers of the electromagnetic force.

Many physicists add a fourth property.

- Photons travel through (the fabric of) space with particulate-wave motion.

** A gauge boson is a "massless" particle that carries one of the fundamental interactions (forces).*

The Motron Hypothesis posits that a photon is a "q-pon," a warpon with quantum mass, (see below) that travels through the 3-Dimensional fabric of space in a linear analogy to the way a stylized ocean wave travels across the surface of the ocean.

The energy of ocean waves travels from one molecule to the next. The molecules of the ocean only travel up and down (or around in a relatively small vertical circle near the surface) as the energy passes.

This is the definitive behavior of wave motion. We might reference this version of wave motion as, "the notion of ocean motion."

MH also posits that the vacuum of space through which photons travel contains a fabric of "dark energy" particles that will be fittingly called "fabrons" that serves as the "medium of passage."** Photons carry electromagnetic energy, whereas ocean waves carry kinetic energy as well as potential energy. The primary dissimilarity between ocean waves and light waves is that the particles of ocean waves are molecules of water (matter) whereas light particles are quanta of light (energy).

** "Medium of passage" will soon be discussed in detail."

MH thus posits the **narrative** that dark energy provides the "ocean" through which photons (and other energy) travel in waves. Therefore, the dark energy being discussed would be the fabric of space, which will be called "the space network." Individual particle-waves of energy that comprise the fabric of space are being called "fabrons."

The need for space to have some kind of configuration called "the space network" will be discussed in "***Relativity Concept***." Many cosmologists make an analogous assumption as revealed by their common usage of the term "fabric of space." The viewpoint provided by MH is a primary upgrade since cosmologists generally think of their "fabric" as "vacuum energy." More on this later.

MH will add one additional facet to this assumption. No longer will we assume that the fabric of space is a passive medium. Depending on its motivation, the idea that the fabric of space is capable of expanding autonomously (under appropriate stimulation) seems reasonable but unprovable at this time.

We have known since 1912 that the universe is expanding. Please note the intentional distinctions between the notion of universal "momentum" expansion and the "motivated" expansion of the space network. The word "motivated" here implies "expanding for a physical reason." This reason will be discussed shortly.

When objects travel through space, it is currently impossible to tell how much of the motion is through, how much is with, and how much is by the supposed network. This one upgrade may be little more than semantics, but it leads ultimately to some of the innovations in this book.

Nota Bene (note well), this upgrade also provides an escape from a glaring contradiction in current theory.

Most cosmologists agree that during the epoch of inflation, the universe expanded faster than light speed c because expansion of the space network is not restricted by the "internal laws" (such as constant c) controlling space motion.

Yet, the space network is supposedly restricted by the "internal laws" (such as momentum) controlling space motion.

Are we going to have a problem here?

Solution. Now we will see that the expansion of the universe was not caused by "momentum" from the supposed "Big Bang." We will shortly be discussing the "motivation" that provided for autonomous expansion of the space network. Note: it was not momentum and there was no "big bang."

Waiver

Of course, physicists of the past cannot be held responsible for any oversite in this regard. Although the existence of a space network had been suspected, its nature had not been revealed until now (with the publication of this book). This revelation will provide the first step in the desperately needed **narrative** for "cosmic acceleration."

Fabrons

Fabrons may "just" be space energy (often called "dark energy"). Even so, they play a crucial role in the narrative of the universe. It is important to ponder the essence of these particles of energy.

Fabrons are probably the only true bosons if they have no mass, so we will call them "null-pons." This will soon be discussed further. Fabrons will be assumed to have these six qualities.

1. They have no charge.

2. They are generally in the form of singularities.

3. Mass is not possible due to their singularity structure.

4. They are highly compressed by universal acceleration.

5. Free, individual motion is not possible due to their compression.

6. They provide all motion and expansion via the space network.

Proof of the existence of a "space network" is rapidly accumulating. This includes the discovery of dark energy, cosmic acceleration, and possibly even gravitation "waves." The consequences of it lead naturally to better understanding of much of the ***narrative*** physics that will be explored in this book. Since many of these things ***might*** exist (in some form) even without a space network, it is important that we spotlight the simplicity reaped and the answers afforded by adopting the <u>*network*</u> version of space.

There might be some question as to the "stability" (consistent over time) of the proposed fabrons. The understanding that will be assumed in this regard will be that fabrons continue to fill the entire universe while providing it with fixed density even as it expands. (This is not as impossible as it sounds … you'll see.)

This stability tends to "anchor" the fabrons in place, not because space has boundaries, but because it is so large and spans so many (look-back) "time zones." In fact, the extent of the space network at any time <u>defines</u> the extent of the universe at that time. The possible boundaries of space will be discussed later.

The filling of "space gaps" and the uniform distribution of fabrons throughout the universe will be assumed to be caused by "photon pressure" that will be discussed in "***Expansion Concept***."

Q: If fabrons are particles of energy, then how do they form a <u>stationary</u> network?

A: It is true that most particles of energy are unable to remain steady (stationary). In "***Ruffle Concept***," we will discuss the <u>**reason**</u> photons (and other monopons) are unable to slow or remain stationary within the space network.

However, since fabrons <u>are</u> the space network, not "<u>within</u>" the space network, then this ***narrative*** does not directly apply to fabrons. This "stationary" property of fabrons will be called "steadiness." Monopons will be discussed in "***Mass Concept***."

Stability vs Steadiness

Providing a mechanism that allows dark energy (fabrons) to become a "stable" platform for all motion somewhat reduces the complexity of space. Stability of fabrons will refer to their ability to maintain fixed density in addition to their steadiness while expanding.

Motron stability will be discussed in "Motron Stability Conjecture" soon. As mentioned earlier, the simpler point of view is generally superior. This is Occam's razor in action.

Singularities

Many theories of physics employ the concept of a "degenerate particle" of some type to explain or absorb the boundary conditions of the theory. Such a particle is often called a "singularity." It is generally believed that the cosmos began as a singularity of energy with "virtually unlimited" or "infinite" energy in a "minuscule" or "infinitesimal" space. An object with such nebulous properties affords tremendous latitude in its realization.

MU will embrace the probability of such a "singularity" in which the quantity of energy (fabrons per unit volume) is referred to as its "density" and attempt to determine some of its other properties based on the discoveries of this theory. These singularity properties will be revealed at appropriate locations throughout this book. This has probably not been done this way before, so it will seem awkward at first. **MU** will also investigate a number of "singularity conjectures" raised for consideration as possible solutions to particular cosmic mysteries.

Timely Plea from the Author

It is likely that many readers will experience the same credibility issues with the characteristics being assigned to invisible objects and forces that also concerned me in the early days of development of this book. I hope that most of these issues will be settled for you within the next 45 to 50 pages.

Please give these concepts time to deploy before dismissing them out of hand. These issues began to achieve traction in my own mind when the "anglewarp" of precession was first considered.

Forward References

You will soon notice that nearly everything in this book is connected to everything else in multiple ways. For this reason, you will occasionally find references to topics that have not yet been covered. It is important to make these forward references for the benefit of those readers who read the book more than once. (To them I say, "Bless you all! May your families increase and prosper.")

("Anglewarp" will be explained soon in "***Angular Momentum Concept.***") Also, please consider the Author's Note when you get there.

Imports of the Motron Hypothesis

There are many happy consequences of the Motron Hypothesis. These, "happy taps" will be revealed as needed. One of the unexpected consequences is the seeming establishment of absolute motion. This might contradict what we have come to believe.

However, any seeming deviation from "established" physics will be adequately explained and justified with solid narratives and strong heart. Supposed deviations will generally be more semantical and less "deviant" than they first seemed.

Absolute Motion

Absolute motion implied by the space network is mostly harmless. For example, because the space network is not perceptible no absolute motion is by it measurable.

However, the notion of absolute motion can simplify the language sometimes. For example, the term "stationary" will now require little qualification when it refers to lack of motion with respect to the presumptive fabric of space.

The concept of relative motion will still be available in a qualified setting. This is normal, however, since reference to a particular frame is such a qualification. Parochial motion within an "*assumed* stationary" frame, for example" should be considered ***absolute*** motion. However, alien motion within that same frame would be considered ***relative***. There will be more discussion with examples of these concepts later.

Whether space (as a whole) is moving or stationary seems to be both relative and irrelevant within this discussion. This will be seen to be compatible with everything else to be discussed in this book.

To deal with relative motion, we only require that "parochial" (local) motion be also considered absolute. All relative motion then becomes absolute (for the purposes of calculations) *within* its frame of reference.

However, alien (non-parochial) motion is always relative unless otherwise indicated to the contrary.

For simplicity, **MU** will posit that the space network is expanding in every direction. If there were a central point in the universe, away-from-which every other point were traveling, it would be indistinguishable, by residents of the universe, from any other point. This is so because the same thing can be said about the motion of every point in space except for points that might be perceived as being (or near) boundary points.

Since the space network is not now perceptible, any parochial location can be presumed stationary at least until the space network or absolute motion someday becomes detectable. Even then, the presumption could continue for relative motion.

Motion within the space network is decidedly different from motion of the space network. However, these motions are not physically discernable.

Therefore, any reference point and any viable coordinate system can be used for measuring movement and distance. The origin and coordinate axes of any coordinate system generally will be considered stationary for the purposes of the study. Thus, all pertinent motion can be understood and specified ***relative*** to these coordinates.

Many systems have advantages and disadvantages that can be used in determining which systems are preferable in particular settings. Mathematicians as well as physicists will continue to use any convenient location for the origin of their preferred coordinate system.

Rather than limiting options, the hypothetical possibility of an "origin-specific" (though expanding) coordinate system allows additional options for relative motion.

Warpons

MU also posits the existence of waves of energy similar to fabrons except that they accompany matter wherever it goes. They move how it moves, and stay where it stays. These invisible waves of energy will be called "warpons" because of their similarity to their implied equivalent in the old concept of "space warp."

A field or span of warpons will be called a "warp field." Warpons and fabrons collectively will be called "motrons" for the role they play in motion. Although motrons are not visible directly, experiments can be performed that make their presence detectible. This will be discussed soon.

It is likely that warp fields and energy waves have already been detected in space. However, these sightings have yet to be identified for precisely what they are. Many energy fields and even small objects are not visible. Nonetheless, we often know they are there due to their influence on other fields or objects.

Consider this question, "How do we know that we are regularly surrounded by invisible and inaudible radio waves?" A radio is required to "convert" the electromagnetic energy of radio waves to audible sound waves.

Warpons are probably "q-pons" with these five qualities.

1. They have no charge.

2. They are incorporated within and surrounding all massive objects.

3. They provide mass to themselves and their associated object. (The ***narrative*** for this will be provided later.)

4. They provide the various warp fields of space.

5. They are the carriers of motion warp (narrative to follow):

    a) *Afterwarp* is responsible for linear motion and linear momentum;

    b) *Anglewarp* is responsible for angular (rotational) motion and angular momentum;

    c) *Orbitwarp* is responsible for orbitar (revolutionary) motion and orbitar momentum;

    d) *Pressurewarp* is responsible for the acceleration of gravity.

Afterwarp, anglewarp, orbitwarp, and pressurewarp will be discussed in detail in "***Ruffle Concept***." (If it turns out that photons, too, are accompanied in their travels by motrons, then those traveling motrons, by definition, are special warpons; these will be called monopons.) There will be much discussion of monopons later.

The generic name for afterwarp, anglewarp and orbitwarp might be "primewarp."

Inertia

Fabrons can be used to clarify and fortify the realm of regal science called "inertia." We think we understand inertia because Sir Isaac Newton "explained" it with his first law of motion, which states: An object at rest tends to remain at rest, while a moving object tends to continue with the same speed (both magnitude and direction) unless acted upon by some external force.

The tendency to remain at rest is called "inertia." The general notion of momentum, in reference to Newton's 1st law, sometimes includes "motion" with zero speed.

How Does Momentum Work?

In this document, for precision, "momentum" will be defined as, "the tendency to continue moving with constant, <u>nonzero</u> speed with respect to any stationary frame.

True, Newton's laws explain what momentum does. However, they do not provide the ***narrative*** of how it works. Using the concepts of fabrons and warpons, we may finally be able to start that conversation.

There are at least four possible mechanisms by which it might conceivably be possible for warpons and fabrons to work together to provide and control motion.

1. Imagine the impulse-motion provided by a warpon gliding through the fabrons of the space network like a snake in the grass.

2. A somewhat similar but far less disturbing analogy is provided by Electric Football® by Tudor Games®.

3. Reality could lie somewhere between the analogies of 1 & 2.

(Back before video games were popular, there was an electric football game (that is still available) in which the board, when turned on, would vibrate causing the player pieces to travel across the board according to their construction based on the type of player represented by the piece and the forward angle at which each was placed on the board.)

The Motron Hypothesis will posit that all motron motion is probably similar to the current concept of photon motion. In addition, this is probably not a coincidence.

Q: ***Why*** is it necessary to have a reason for momentum? Why not just continue to assume "momentum is natural and force is required to alter it?"

Comment: We will return to this question shortly. First, however, let us use this question (and the way it is worded) to illustrate a very important point.

Physics 101

Never allow the intrusion of the word "why" into a physical inquiry to allow you to sidetrack the responsibility to provide the required physical narratives for **all** physical procedures in addition to the mathematical formulas that might detail their behavior.

A: This confusing question illustrates why science sometimes refuses to deal with "why" questions. To answer this question we must first understand the question, which was, "Why is it necessary to have a reason for momentum?"

Some might wonder, "What causes an object in motion to continue in motion unless acted upon by some external force?" Instead, could we not wonder, "What else could it do?"

This last question implies that, in our understanding of the universe, there is no clear alternative but for it to continue with the same motion. This lack of understanding is exactly the point and this returns us to the question of *how the universe works*, in this case … how momentum works.

Now we see that this "Q" need not "really" be a why question. We can only hope that the Motron Hypothesis above will finally allow us to answer this question.

Requirements for Phenomena

There are at least two levels of requirements within this presentation.

It sometimes requires a ***narrative*** to explain the reasons (requirements) that something behaves the way that it does. Some of the more obvious requirements discussed in this book are often not labeled as ***narratives***.

Here are two sample problems. To meet the requirements presented here, the details that must happen are:

1. To evaluate our understanding, we might try to decide if photon pressure seems sufficient to explain universal expansion.

2. Similarly, we might try to decide if the need for constant "parochial c" is dependent enough to require also that "alien c" be constant. This important detail will be discussed at length.

To answer these questions might require us to dig a little deeper. This book will try to help you 1) find the right answers in addition to 2) "how" to find them. However, you must first ask yourself the right questions. (Professionals already know most of the questions.)

1st Uni-Verse

"It is critically important,
That present understanding,
Simplify past knowledge,
So that future vison
Can become clearer."

2. Inertia and Momentum Concepts

Introduction

Do not skip this concept. It is a prerequisite for any understanding of gravity. This concept assumes a familiarity with the basic facts of inertia and momentum as established by Newton. We will attempt to explain how the space network supports inertia and momentum.

However, even those who are less familiar with inertia and momentum will be provided with some insight into these concepts. It might be advisable to keep your browser handy (just in case you need it) as you read this book.

A glossary is provided in the back of this book for terms that cannot be looked up anywhere else. Many of these terms are new. Some are old words that have been assigned new meanings, interpretations, or some broader understanding. Any word with a changed definition will be considered "new vocabulary."

One other word of advice for young scholars. As you read this book, you will likely have many questions. This is a very good thing.

Simple things like definitions should be looked up as you go to prevent a "logjam." However, a notebook (paper or electronic) should be kept handy to keep track of your personal thoughts and curiosities. This will allow you to stay on point as you study this material. You would then not need to be concerned about what is important and what you will need to remember.

It is very possible that you will have questions that go beyond the scope of this book. However, most of your questions will likely require preparation before they can be answered. Do not be satisfied until all of your questions are eventually answered to your satisfaction.

Warp Fields

If the motion of any object (with its accompanying warp field) is maintained by any impulse (such as an electromagnetic, manual, gravitational, or even an explosive impulse) the warp (energy) field of that object will continue to travel with and around the object, but will lag (gather or concentrate) behind to maintain an ellipsoidal "afterwarp."

This afterwarp of "dark matter" will comprise an invisible component of the object. It can (currently) be detected only by its effects on moving objects. By its nature, the afterwarp would appear to be caused by inertia. On the contrary, inertia is caused by and is a measure of the warp lag, which is the cause of momentum. The causes and consequences of this ellipsoidal afterwarp will be discussed in the pages that follow immediately.

Before any force displaces (moves) a stationary object, the three-dimensional warp field is symmetrically distributed throughout the universe, around every component (point) of the object, with its influence on the component diminishing as the square of the distance from the center of the component. This "inverse square law" is mindful of the relation between the intensity of a light beam and the distance from the source.

Because of this similarity, it is likely that individual warpons align themselves "spherically," not like a web, but in linear "waves." We can think of the waves as if they are radiating in 3 dimensions away from the center of the sphere (essentially to infinity).

This is somewhat akin to light rays radiating away from the sun in a painting. In this case, the sun becomes the center of gravity of the component. Due to their effect on the object, however, it is best to think of the warp field as radiating in the direction of the wave motion toward the center as if pushing the component from every 3-Dimensional direction. The physical strength that maintains the symmetry is the source of the inertia.

When an object is moved in any direction by any force, "disturbing its warp symmetry," the warp field is compelled to adjust. Symmetry is actually a mathematical property (label) with no physical consequences. Symmetry is merely an indicator of current conditions, which themselves <u>may</u> have consequences. The rending action between the applied force and the resistance of the surrounding fields (warpons and fabrons) cause the symmetrical modifications.

Again, the stability of the warp field is not due to inertia, but to the connections and interactions between the space network and each component of the "massive object" (an object possessing measurable mass) associated with the warp field.

Minimum Impulse

As mentioned above, many impulses can affect the motion of an object. These include general impulse like manual and mechanical impulses. Friction and structural forces are only considered impulses, however, when they have a stated or implied time factor. If insufficient unbalanced impulse is applied to cause motion, then the warp-field symmetry will not change.

Thus, an external applied impulse does not cause the afterwarp directly. However, sufficient unbalanced impulse over time (impulse = force · time) can cause motion that instantly creates an afterwarp that maintains the speed. ***Momentum is the ability to keep moving after the original impulse to move has ceased.***

Since force is equal to mass times acceleration and acceleration has a time component (acceleration equals distance divided by time squared), then mass times speed can be called impulse (J). Therefore, Newton's 2nd law implies that impulse: $J = m \cdot v = m \cdot d / t = F \cdot d / v$.

Section Headings

Note: all headings that are centered and underlined (as above) presage the inception of advanced details. These sections can be skipped with little consequence by neophytes who are less concerned with details than with basic concepts.

Connoisseurs with refined focus might also skip marked sections unrelated to their focus. The advanced details (thus labeled) relate primarily to ramification, justification and verification rather than the substance of **MU**. These details, being clearly marked, allow readers to return for these enhancements once the basics of this work are sufficiently taken in.

Warp Fields (Continued)

The concept of "Warp Field" in **MU** replaces the concept of a supposed "space warp"; it should not be thought of in terms of a distortion of the geometry of space, but rather in terms of a (much smaller) "presence" of warpons within the space network of fabrons.

Geometry becomes much too complicated if we consider portions of it to be somehow "warped." The seeming simplicity of "space warp" comes with a high "price" little heart and no necessity.

Warp fields accompany all massive objects and contain dark matter capable of controlling the motion of those objects as well as affecting the motions of other nearby objects.

The maximum contiguous warp field of any object that can affect nearby objects will be considered the "*superfield*" of that object. The portion of the superfield that is physically attached to the object and has the ability to control the objects own motion will be considered a "*subfield*" of that object. The subfield is probably a "stronger" field than the corresponding superfield. However, the subfield is responsible for speed while the superfield primarily causes acceleration.

"Afterwarp" can be caused by any of the "warp displacement" conditions listed next. "Space warp" of general relativity will be discussed later in "***Mass Concept***."

Equivalent Mass

The "warp rate" of any moving object is a measure of the relative loss of symmetry of its associated warp field. The "warp value" associated with any object will refer to the quantity of "equivalent mass" that is currently thought provided to that object (through its warp) by its speed. The concept of equivalent mass will be discussed in "***Relativity Concept***." The source of mass (m) will be discussed in "***Mass Concept***."

Before the source of mass was understood, equivalent mass (m_e) was believed accrued by a moving object with rest mass (m_0) and acceleration (a) depending on its speed (v), by the formula:

$F = ma = (m_0 / \sqrt{1-(v/c)^2}\,) \cdot a = (m_0 + m_e) \cdot a$, so that $m_e = m - m_0 = m_0 / \sqrt{1-(v/c)^2} - m_0 = m_0 \cdot [(1 - v^2/c^2)^{-1/2} - 1]$.

A stationary object has a warp rate of 0% and a warp value of 0. A warp value of 1 corresponds to a warp rate of 100% warp displacement as manifested by photons, which travel with maximum speed, c. The warp value for moving particles will be counted in "warp units" (wu) between a minimum of 0 wu and a maximum of 1 wu.

Warp values between 0 and 1 would normally be expressed in decimal form such as 0.0123 wu, for example. The warp value of large moving objects can (as needed) be measured in mass units such as micrograms (μg), milligrams (mg), grams (g), or kilograms (kg) … etc. to reflect their "equivalent mass" as it was previously envisioned.

In this respect, the FitzGerald quotient factor $\sqrt{1-(v/c)^2}$ will be called the "force-effectiveness index" or simply the "force index." This quotient factor represents the quotient to f that is required to make effective force $F = f / \sqrt{1-(v/c)^2} = m \cdot a$, where "f" is the "baseline" force, "v" is the speed of the object, and c is light speed. Baseline force will be discussed in "***Relativity Concept***" near the end of this book.

Note that when $v = 0$, that F = f. Furthermore, since the converse of this conversion is also true, then only when at rest is mass (m) a true proportionality constant between baseline force (f) and acceleration (a). I.e. force (f) and acceleration (a) are not truly proportional except when the object is stationary.

However, when stationary within any system, there is no speed with respect to that system. Due to the biconditional nature of the above relation, then this relation creates the *new definition* for “effective force F” as stated above.

Afterwarp

If an unbalanced force is sufficient to overcome the inertia of an object and cause the object to accelerate in the direction of the force (vector), the motion will cause an (invisible) aft (trailing the moving object) displacement of the warp field. This displaced warp field will be called the “afterwarp” of the object. This afterwarp will physically maintain the motion.

The name “afterwarp” is somewhat mindful of a cross between the afterburners of a jet engine and the “warp drive” of a “star ship.” As long as sufficient force is applied to a massive object, the object will accelerate. This acceleration will continually increase the (invisible) eccentricity of the ellipsoidal afterwarp.

Any force that is applied for any length of time is called “impulse.” When the impulse ends, the object continues its motion with constant speed until acted upon (again) by an external impulse. This is Newton’s first law of motion.

Newton’s second law tells us that force F must be equal to the mass m of the object times the quantity of desired acceleration a, $F = m \cdot a$. It should be emphasized that disparately, force is associated with acceleration while impulse is associated with speed. Other implications of these laws will be considered later.

As long as an object continues its motion, its afterwarp cannot regain its symmetry. (We can mentally imagine that the attempt to regain symmetry is what maintains the motion.)

This is somewhat analogous to the "impulse" for a physically off-balanced person to run headlong in a particular direction in an attempt to keep from falling in that direction. Thus, uniform motion maintains the eccentricity of the afterwarp that, in turn, maintains the uniform motion.

This **narrative #1** finally explains (and visualizes) inertia and momentum of Newton's first law and removes them from the "realm of regality." (More explanations and verifications follow shortly.)

Notice that the above discussion assumes a uniform warp field to attain symmetry. If the warp field of a particular object were not symmetrical with respect to the space network, the asymmetry would define its required (absolute) speed of travel through space to maintain the eccentricity. However, symmetry and eccentricity as well as the space network and absolute motion are not currently visible.

Vehicular Example One

Consider an example where a fire engine is rolling out of the fire station on a flat horizontal surface. Due to the structure of the fire engine, forward motion is uniform for the entire truck. This forward motion maintains the ruffling of the warpons (called afterwarp) behind every part of the truck. This afterwarp maintains speed and the truck rolls uniformly forward.

The field of warpons around any *stationary* object is symmetrical. As mentioned above, it requires some type of minimum unbalanced force to displace it.

The more massive an object is the greater is its warp field. The greater the field, the more steadfast its motion. Specifically, the force required to displace the field of a stationary object is directly proportional to the mass of the object.

The visible components of this motion can be (and have been) verified empirically (experimentally).

If the force continues to be applied to the object, then normally* the field displacement will continue increasing proportionately. The greater the applied force or the longer it is applied, the greater the resulting acceleration.

Thus, force f in the appropriate units is normally* equal to the quantity of mass m multiplied by the acceleration a. This is Newton's 2nd law of motion, $f = m \cdot a$. This **narrative #2** explains acceleration of Newton's second law and removes it from the realm of regality.

* *We will discuss relativistic velocities in* **"Relativity Concept."**

Historical Note

There is no physical significance to the use of fire engines for examples. The first time these concepts were presented to students in a seminar on relativity in May 1997, the author borrowed a small fire engine from his children's toy box that was just the right size for use in illustrations of momentum, gravity, and relativity.

A fire engine along with fire hoses and firefighters were symbolically included in that early version of **MU** (then called MOTION THEORY) which was (finally) being corralled and encapsulated into print (though never published).

In a nod to history, those illustrations are similarly being physically written into this new and greatly expanded version of **MU**.

Vehicular Example Two

Consider another example where a fire engine is leaving the station. As power from the engine is transmitted to the wheels through the drive shaft and axle, the wheels try to rotate. Forward rotation of the wheels applies a backward force against the ground. The force of the rotating wheels against the ground is matched by an equal and opposite "***linear reaction***" of the ground against the wheels. This is an illustration of "Newton's 3rd law."

There are, of course, many mathematical equations that physicists use to express the precise relations above. However, in this book, we will be primarily interested in the new narratives expressing the causes rather than the formulas that express the well-known effects.

Other Aspects of Motion

When the situation is reversed and a linear motion causes an object to rotate, the reaction will be called an "angular reaction." An example of angular reaction would occur when a child uses a stick to roll a hoop along level ground.

Suppose object X and object Y are in contact and x is the force of X on Y while y is the force of Y on X and no other forces were affecting the motion of either. If we consider X and Y to be the two parts of a two part object called "XY" that is traveling with constant speed, the warp field of XY would be the accumulation of both fields just as all warp fields are accumulations of their subfields.

Then the warp displacement on XY caused by x and the warp displacement on XY caused by y will balance symmetrically only if the two forces x and y are equal and opposite since both forces are acting on the same object XY.

If the warp displacement on XY were not balanced, then XY would accelerate according to Newton's second law. In that case, XY would not be traveling with constant speed as specified.

The linear effect of object X upon object Y is also called "linear reaction" or simply "reaction" when the meaning is clear. Any rotational effect of object X upon object Y would be called "angular reaction." It should be noted that a rotational action could cause linear reaction, angular reaction, or both.

Even before this discussion, both of these reactions were well documented and well understood. This completes *narrative #3*, which now explains how Newton's third law works and removes it from the realm of regality.

If the force on the truck provided by the forward rotating force of the wheels against the ground were not matched by some equal and opposite force upon the truck, then we would have an unbalanced force. Newton's second law tells us that unbalanced forces cause acceleration.

If the wheels do not slip, and (God forbid) the firetruck does not self-destruct or fall apart and the ground does not give way, then the only way to accommodate this unbalanced forward rotating force of the wheels against the ground is for the truck to move forward.

This initial forward motion of the truck causes an increase in the afterwarp on the truck. This, in turn, maintains the impulse-motion (speed).

The afterwarp increases as the motion increases. This increase causes the fire engine to accelerate forward.

If the force ceases, the truck continues to move forward with fixed speed due to the conformity of the fixed afterwarp. Constant speed will continue until the afterwarp is altered by the application of new or renewed force. This completes *narratives numbered 1, 2, & 3* concerning Newton's laws.

Imports of "*Momentum Concept*"

These concept *narratives* not only simplify our understanding of motion, they also simplify the *narrative* necessary to explain related topics. In the process of this simplification, we have also removed all three of Newton's laws of motion from the realm of regality.

Coming Soon

Probably the most serious consequence of "***Momentum Concept***" is that momentum can no longer be considered something that "just behaves that way." We have long known that force causes acceleration and other motion. When the force ends then the acceleration ends also. Then what?

We know that momentum continues even after the force ends. Now we know that something causes momentum and angular momentum. So … what is it?

Since force causes acceleration and acceleration increases speed, then force indirectly increases speed. Increased speed, in turn, increases afterwarp, which is responsible for momentum. Similarly, anglewarp is responsible for angular momentum (that is generally assumed to include orbitar momentum.)

Furthermore, afterwarp, anglewarp and orbitwarp are responsible for linear, angular and orbitar pressurewarps respectively, which are responsible for gravitation, precession and nielucion respectively. These are the concepts that have been missing. This will all be explained very soon.

Thus, momentum, angular momentum, orbitar momentum, gravitation, precession, and nielucion all have warp as a cause and force as a primary factor. This makes them all substantive force products with "warp" parents. (No, not warped parents.)

Classification of Linear Motion

Gravitation is related to its relative proximity to the source of the gravitation; furthermore, since it causes acceleration, gravitation is classed as a force. Momentum, angular momentum, and orbitar momentum require auxiliary sources of acceleration, but all result in continuation of speed. This means that momentum, angular momentum, and orbitar momentum are all caused by impulses.

It might be useful to classify motion properties as they are examined individually. Momentum of Newton's 1st law of motion will be classified as a "direct primary action" because a force acting on an object causes these actions directly. Linear reaction of Newton's 3rd law will be classified as a "direct secondary reaction" since an action of one object is directly causing a reaction in a second object.

Classification Tables

Linear Warp Unification Table 1	**Primary Action**	**Secondary Reaction**
Direct Cause (Afterwarp)	**Momentum**	**Rebound**
Indirect Cause (Pressurewarp)	**Gravitation**	**Gravity Assist**

The next two paragraphs are included to complete Table 1. These motions will be discussed later.

Motion caused by gravitation will be classified as "indirect primary action" whenever any ruffling of the afterwarp of an object indirectly distorts the motion of the object.

Motion caused by gravity assist will be classified as an "indirect secondary reaction" because the warp field of one object indirectly creates a pressurewarp on a nearby object, which "induces" acceleration of the nearby object. In the case of linear motion, the acceleration is called gravity assist.

Cross classifications of properties often lead to new discoveries. The creation of the "Periodic Table of the Elements," the "Standard Model of Elementary Particles," the "Hertzsprung-Russell (H-R) Diagram of stars, etc. have all led to new discoveries and deeper understanding of the subject of the classifications.

The “Linear Warp Unification Table” here and the “Angular Warp Unification Table” in the next chapter both had gaps when first established for the furtherance of this presentation. This suggested a potential for new discovery. The eventual discovery of examples of these actions has greatly enhanced this presentation of linear motion.

2nd Uni-Verse

"Momentum is the ability to keep moving
After the impulse to move has ceased.
One of our strongest personal traits
Must then be our inner momentum."

3. Angular Momentum Concept

Anglewarp

Just as warp is altered by motion, it is also altered by rotation. Instead of being shifted linearly, however, the "anglewarp" is rotated angularly.

For example: if an object is rotating counter-clockwise as seen from above, the near side warp would still be straight, but would be rotating with the object from left to right as seen from any side (if it were visible), causing the rotation to continue.

It is being assumed that the anglewarp rotates with the induced rotation of the object just as the afterwarp always travels with the induced linear motion of its massive object. Both of these details are revealed by their consequences; these will be investigated.

This ***narrative #4*** of the anglewarp explains the phenomenon called "angular momentum" removing it from the realm of regality.

Thinking that the anglewarp cannot be "seen" does not mean that its presence cannot be demonstrated.

Decreasing the Radius

We know that the conservation of angular momentum is similar to the conservation of linear momentum. Therefore, using the formula for angular momentum $L = r \cdot m \cdot v \sin \theta$ (where v is the linear speed and θ is the angle of rotation) might allow us to predict how decreasing the radius r would increase the angular speed $\omega = \frac{v \sin \theta}{}$

even if the other parameters (mass m and angular momentum L) were unknown (but constant).

However, understanding the ***narrative*** is a little more challenging and much more rewarding than just understanding the definition and the formula. Try this mental experiment.

Rotation Experiment

Imagine trying to turn an adjustable crank to rotate a machine or a heavy wheel (such as a crank-start engine or the grinding wheel that we will consider now). Initial trials easily demonstrate that a longer handle makes the crank easier to rotate.

Once the wheel has established sufficient angular momentum, however, the crank cannot turn any harder than the rotational resistance. With accelerated cranking, the resistance could drop to near zero as a result of its rapid rotation rate. (Just enough to maintain current rotation.)

If the length of the crank were then shortened, the applied rotation force will be closer to the axis and will suddenly experience resistance. This allows (causes) the crank to turn harder and thus rotate the wheel faster. This, then, increases the angular speed proportionately.

This completes the ***narrative #5*** understanding of the relation between the angular speed $\omega = v \cdot \sin\theta$ and radius r, as they are involved in "angular momentum L," removing it from the realm of regality.

Terminology

To avoid confusion, it will be necessary to define several terms specifically created for **MU**. These are arranged in logical order rather than alphabetical order.

1. Motrons are waves of dark energy that fill the universe and are responsible for all motion. The two types of motrons are called "fabrons" and "warpons."

2. Fabrons are motrons that provide the fabric of space called the "space network."

3. Warpons are motrons that accompany matter, move with it and stay wherever it stays. Warpons are named for their conceptual role in "warping space" for those who have learned to express energy fields in those words.

4. The portion of the parochial space network (through which a massive object and its warpons are traveling) is historically called its "frame of reference" or simply its "frame."

5. A warp field is a field or span of warpons.

6. Afterwarp is a warp field generated by linear motion of an object within its frame. The afterwarp is responsible for the linear momentum of the object and also for gravitation with any nearby object.

7. Anglewarp is a warp field generated by the rotational motion of an object within its frame. The anglewarp is responsible for the angular momentum of the object and also for precession of any rotation in the presence of any nearby gravitational field.

8. Spinwarp is a specific anglewarp caused by the effect of a rotational motion (spin) of an object.

9. Vectorwarp is a compound warp caused by two or more simultaneous warp sources on the same one or more object(s).

10. Spiralwarp is a vectorwarp, generated by a continuous spinwarp progression, about a center that is moving along a path "parallel" to the direction of its simultaneous afterwarp.

11. Orbitwarp is a warp field generated by the revolutionary motion of an object about a central massive object. The orbitwarp is responsible for the orbitar momentum of an object and also for nielucion about any rotating central object.

12. Pressurewarp is a distortion of one warp field caused by the proximity of a second warp field.

Rotational Stability

The presence of spinwarp (all the way around a rotating object) provides rotational stability to the object. Sufficient spinwarp can maintain a vertical rotating frame such as a spinning top. This is similar to the stability provided by angular supports all along a bridge, gate, bookcase, etc. A football and a bullet are prevented from tumbling through the air (in any direction) by giving them a little angular momentum. This angular support ***narrative #6*** explains rotational stability removing it from the realm of regality.

For any projectile, if the axis of rotation were parallel to the direction of travel, this would make it tangent to the arc of travel when gravity is involved. To reduce wind resistance, it is best that the axis of rotation be the "major axis," or the axis with greatest length of the object.

Precession

The existence of spinwarp can actually be demonstrated. When a gyroscope is spinning, it may be possible to force the axis of rotation into the spinwarp by tipping the axis. Keep in mind that the invisible spinwarp is causing the gyroscope to continue spinning on its axis.

When gravitation or any other external force tilts the axis of the gyroscope into the path of the spinwarp (that is always rotating around the spinning object), because of its rotational stability, the axis tends to visibly "bounce off" or "rotate with" (usually both) the spinwarp supports (such as the gimbals of a gyroscope).

This description is probably oversimplified, but since the spinwarp is not visible, then the visible reaction is our primary physical evidence of its existence. The action of the invisible spinwarp exposes its presence.

The spinwarp was conveniently out of the way while the rotating warpons kept the gyroscope spinning around its axis. Tilting the axis in any direction causes an encounter between the spinning warp and the physical axis. This causes the axis to rotate (precess) awkwardly in the direction of the slanting spinwarp.

This trace of independence of the gyroscope from the spinwarp illustrates that spinwarp is a property of the warp field and not of the gyroscope. The initial bounce of the precession can cause a "nodding" oscillation called a "nutation."

The "rotation" of the axis illustrates the well-known phenomenon called "precession." This rotation is actually called "precession of the axis."

To differentiate (verbally) between (a) the rotation of an object around its axis and (b) the precession of its axis, the axis around which "the axis of rotation" precesses, is called the "axis of precession." The orientation of the axis of precession and the speed of precession are both affected by the speed of rotation, the strength of the precession force, and the direction of rotation with respect to the source of the precession force.

The cause of precession will be presented soon. The direction of precession can be remembered by the "right-hand rule" mnemonic. Using your right hand, wrap your four fingers as if around the axis of the rotating body with your fingers curved in the direction of the rotation. The polar direction of your right thumb will be labeled "N."

With your right hand still in place, rotate your right hand around the axis until your straightened right index finger points in the direction that the external force is pushing the N-pole. Your other three fingers (that are now perpendicular to your pointing index finger) will then indicate the direction of the resulting precession around the axis cuddled in your hand.

Once the hidden forces are understood, it can be "seen" how the tilt into the anglewarp causes the precession. Understanding the physics of **narrative #7**, concerning precession, <u>requires</u> a physical concept such as spinwarp; this finally explains (beyond the mnemonic) <u>how</u> precession works and removes it from the <u>realm of regality</u>.

The use of afterwarp or other motion to alter the spinwarp (in any way) creates vectorwarp. Mathematics can be used to model (describe) precession in terms of vectors. If all one cares about is to know what precession does, how much, and in what direction, then vector analysis is a wonderful convention.

Distortion of an object's warp field by rotation (spin) of the object produces spinwarp. Combining two warp forces (such as afterwarp with spinwarp) will be called "vectorwarp." Two or more separate warp fields can occur simultaneously. The most common vectorwarp can be thought of in terms of two perpendicular components. A component tangent to a circular rotation can be considered spinwarp. A component perpendicular to a plane of circular rotation can be considered afterwarp. At any moment, vectorwarp can be thought of as the vector sum of its components.

Inverse Precession

Due to the physical relationship between rotation and precession, it is also possible to reverse the relationship. By manually increasing the precession, it is possible to accelerate the rotation. This might be useful as an exercise machine, or any time we wanted to generate or store rotational motion.

For the purpose of this discussion, it is useful to know that precession is reversible. It is doubtful that the reverse process will be called un-precession, however. It could possibly be called "anti-precession" or possibly "inverse precession."

Components of Precession

Precession always has two components, the warp component and the distortion component. The most recognized warp component is the anglewarp. This type of precession is called “angular precession” or simply “precession” when the meaning is clear.

Precession can also occur with afterwarp. When the warp field of precession is afterwarp, the precession can be called “linear precession” or “gravitation.” Since gravitation has not previously been understood, it has not previously been recognized as linear precession.

Finally, precession can occur with orbitwarp. This type of precession can be called “orbitar precession” or “nielucion.” These latter two types of precession will be discussed later.

The second component of precession is the “precession force” or simply the “distortion.” The distortion required for precession can be caused by such things as gravitation, magnetism, moving fluids, or manual distortion. These latter two forms are considered “artificial.” The *second equivalence principle* called “anglewarp equivalence” will be discussed in “***Gravitation Concept***.”

Bernoulli Curveballs

Curveballs in sports are not always caused by precession. The Bernoulli principle can also cause a curveball. Bernoulli’s principle will be discussed later in this chapter.

Spin on the ball introduces anglewarp. The anglewarp can then cause air pressure to build up on one side of the ball depending on the type of spin and the direction and speed of the ball.

Distortion of the surrounding air can be caused or enhanced by such things as the dimples on a golf ball, fuzz on a tennis ball, or the raised stitches on a baseball. These are the necessary conditions to invoke the Bernoulli principle. This will be discussed in "***Expansion Concept.***"

A ping-pong ball is so small and light that very little distortion is needed. It curves considerably with proper paddle motion. If it were slightly larger or heaver the ball might need to be given a rougher surface.

Notice that the delivery of motion to all of the balls mentioned above is designed to provide the required power and spin. Witness the ribbed face of a golf club, the tight strings of a tennis racket, and the sandpaper on a ping-pong paddle.

In all of these cases, the delivery of the ball provides the necessary power and spin. The friction and viscosity of the air or other contact media provide the necessary (Bernoulli) distortion force.

It should be mentioned that a Bernoulli curve is also possible with a football. However, this would require the ball to be delivered *perpendicular* to the direction of travel. This tends to reduce the range of the pass, but distance is not always the most important factor. A Bernoulli pass can cause the ball to float or to drop unexpectedly. This could confound a defender who has never heard of this possibility or seen such a pass delivered. A Bernoulli football pass is not likely to be overly dramatic, however, due to the weight of the ball and the small area of stitches. Of course, there is always the possibility of a new game with a new ball … "Bernoulli ball," anyone?

Precession (Continued)

Precession also requires a spinwarp and a distortion force. With a spinning top, the distortion is caused by earth's gravity acting on the spinning top, which becomes off-balanced as the spinning slows. Earth's precession is caused by the sun's gravity acting on the slight "distortions" that make earth an "oblate spheroid." Precession in sports generally requires a ball that is _*not*_ spherical.

<u>Aligned Spiral Pass</u>

The following example may be of interest to football players and fans. The shape of the football is important to the game because it reduces the chance of a "regular" curveball ("Bernoulli curve") while also making any <u>attempt</u> perceptible in the orientation of the ball. The shape of the football also makes "aligned precession" possible. This is the name we are giving to a spiral pass that maintains the ball's tangential alignment throughout its trajectory.

Aligned precession of a passed football is generally desirable. As the football is released by the passer, the axis of the spiral needs to be slanted upward in the direction and at the precise angle the ball will initially be traveling through the air. The football, as a projectile in still air, will travel a (nearly) parabolic trajectory.

As the direction of travel arcs forward, the axis of rotation (spiral), if not for precession, would lag behind. This would expose the nose of the football to upward wind pressure. For a right-handed passer, upward wind pressure would cause the nose of the ball to try to tip up.

The physical anglewarp is rotating to the right across the top of the ball as it maintains the ball's rotation. If a misalignment between the axis of rotation and the vector of travel causes an encounter between the axis and the anglewarp, it can cause a slight precession to the right as seen by a right-handed passer. For left-handed passers the precession would be similar but to the left.

In either case, the initial shift in wind pressure would simultaneously rotate the precession downward. This is useful when the ball arcs through the air so that the nose of the football will ultimately precess downward as the ball is transcending the top of the arc and starting its downward path.

Notice that from-right-to-down implies clockwise for the right handed passer, while from-left-to-down implies counter-clockwise for the left handed passer. In both cases, this matches the spin the passer is imparting to the football.

As long as the axis of rotation is aligned with the trajectory (and absent strong surface winds), there is no danger of over precession. The downward precession of the axis of the football realigns and maintains the axis of rotation tangent to the flight path of the football, thus eliminating the lateral pressure on the nose of the football. This prevents further precession.

Precession will be kept in check because, as the axis of rotation arches forward and then downward, realignment of the axis with the flight path diminishes the lateral wind pressure on the nose of the football proportionately. This reduces or eliminates the precession and maintains minimum resistance.

Due to precession, all of the reactions listed above occur sequentially so that the axis alignment is maintained throughout the process. Discriminating observers and fans of the game consider this adjusting alignment a magical and a beautiful thing.

Precession of the football is also useful to the passer since reduction of wind resistance increases the range (distance) and accuracy of the pass.

It can also be demonstrated that precession is a reaction to the headwind on the axis of the football and not just another property of rotation. For example, there would be no precession were this experiment to be performed by a robot in a vacuum, or near vacuum such as in space, on the moon, or in an evacuated pressure chamber.

References above to "wind" when passing the ball in *still air* refer to the air "passing over" the *moving ball*.

"Blowing" wind can complicate the situation considerably. Not only can blowing wind alter the ball's trajectory, it can also cause sidewise precession. Players should practice in windy conditions as often as practical even if the "practice" wind must be artificial.

Notice also that air is not directly responsible for precession, but air is necessary for wind, and headwind can provide force and lateral (initially upward) force on the nose is necessary for precession.

Precession of spinwarp is not new science, but its **narrative** does provide better understanding. All football coaches should know this. However, most people seem to think that greater power behind the pass combined with the resistance of the air reduces the precessional "wobble."

It is probably true, however, that greater distance is produced by less wobble. Contrariwise, less wobble is <u>not</u> directly produced by the greater force required for greater speed and greater distance. However, greater speed does increase wind pressure on the nose of the ball, which might reinforce the precession. This can sometimes indirectly reduce wobble depending on the direction of the precession. This type of "converse confusion" is a common logical fallacy.

Note the role (in causing precession) played by the nose of the football. Since most balls are spherical, this explains why precession is not common in other sports.

(It should be noted here that, of course, quarterbacks could gain this skill without knowing about precession. They can just keep practicing until they get the result they want. Since this discussion is already parenthetical, a shout-out might be appropriate to fellow former Baylor Bear RG3 from M^2H2. Beautiful job, <u>RG!!!</u>)

Vehicular Example Three

Consider the fire engine, again. If the driver turns the steering wheel to the left while the truck is in forward motion, (skipping some of the details), a force is transmitted from the pavement, by way of the front wheels, to the entire front end of the truck.

The counter-force (reaction) of the ground on the front wheels initiates the left hand turn of the truck. This turning initiates the anglewarp on the front of the fire engine by rotating the afterwarp to the left if it could be seen from the driver's seat.

These forces cause the fire engine to continue at almost the same speed while rotating counter-clockwise (if seen from above). The anglewarp tries to maintain the rate of rotation until it is again adjusted by the driver.

Most of the reaction felt in the steering wheel as the driver shortens the radius of curvature is not caused by the anglewarp, but by the deviation of travel from linear. The slight loss of speed seems to be caused mostly by increased friction in the moving parts and between the tires and the pavement.

Some loss of speed is caused by the fact that, when cornering, the turn-centers are usually not the same (concentric) for all four wheels, especially the front two (the steerable two). This is partially responsible for the increased wear on the front tires even when both the power axle and the center of gravity are near the rear of the car. However, the front wheels (on most cars) are also responsible for changing the direction of the linear momentum.

[Turn centers do not depend on speed, so they could be considered by the engineers who design steering linkage. This process could be called, "turn-center coordination." This coordination can be addressed, if it is not already, by either proportionately decreasing the turn angle of the outside front wheel or increasing the turn angle of the inside front wheel. Using the inside wheel would likely be more accurate due to the larger adjustment angle caused by the shorter turn radius. Designation of outside vs. inside wheel is dependent only upon which way the steering wheel is turned.]

After completing the turn, additional force is required to stop the vehicular rotation. If the vehicle were to enter an icy patch during the turn maneuver, it could continue its rotation in addition to its linear slide. This phenomenon is uncomfortably familiar to most that often drive in wet or icy conditions.

Note that the primary change caused by the loss of traction is that the <u>curved path has become straight</u> (this default property of the afterwarp will be called the "linear bent" of motion) while the vehicle rotation and forward motion are maintained by angular momentum and linear momentum independently.

Linear Bent

Here is how it works. We know that impulse is required to establish momentum. Once established, however, we imagine that no action is necessary to maintain the "linear bent" of momentum. This is not always true, however.

Whenever a lateral force F is applied to a moving object, the object does not instantly travel in the direction of F. There continues to be a "forward component of momentum."

<u>Momentum of Rotation</u>

This forward component of momentum is being called the "linear bent" of the afterwarp. Precession and nielucion both require that centripetal force be provided by some radial arm such as gravitation to divert the linear bent.

Arching the plane of any curve (such as on a highway or race track) upward beginning on the exterior edge of the pavement tends to decrease the traction required for turn-center coordination and to maintain the curved path.

Traction is needed to overcome the linear bent and maintain the (flat) curved path. Traction is not necessary for linear momentum or for angular momentum. This fact lends credence and understanding to the presence of afterwarp and anglewarp.

If the road surface were arched sufficiently (call this a "speed gradient") so that the steering wheel need not be rotated (while making a turn at a specific speed), then the turn centers would all be simultaneously at infinity since all wheel axes would remain parallel. Under these conditions, traction is primarily important only when changing speed (especially when stopping).

The name "speed gradient," or "speed grade" for short, might be useful to warn future drivers, *for example*, of an "86 mph speed grade ahead." A nice addition would be a parallel stripe or lane for slower drivers (below 86 mph) to stay below or in.

Under these maximum-gradient conditions, the safest path on a race track in a crowded field of high speed drivers would be the outside (upper) edge of the track. Cars slowing down for any reason would tend to migrate downward and inward toward the center of the track.

This would likely result in less rainwater and ice accumulating on the pavement, and also fewer collisions with the outside wall and fewer wheels and parts up into the stands of a race track. Furthermore, upward collisions would tend to be less violent. However, reactions of even professional drivers cannot generally be accurately predicted. Some of these things have been tried with some success.

Anglewarp contributes to the destructive nature of cyclones. Ice skaters learn to alternately use and then compensate for angular momentum in their turns and rotations. Angle warp makes figure skating, football passing, and auto racing much more difficult than they first seem.

A Larger Example of Spinwarp

The rotation of the sun on its axis must be (and is) maintained by spinwarp. This spinwarp has the potential to affect nearby planetary orbital motion within that warp field.

This indirect reaction would naturally be greatest for the inner planets where the spinwarp effect is greatest. This would explain the nielucion of Mercury's elliptical orbit. Nielucion was originally thought to be caused by "space warp." More on this later.

In an orbit revised by nielucion, the major and minor axes of the counter-clockwise elliptical orbits, themselves, rotate counter-clockwise, as they should by this theory, since the sun also rotates counterclockwise.

Note the difference between a revised (rotated) orbitar path like this and a revised orbitar period (timetable) in which the path remains unchanged while the orbiter advances. This distinction could be misdiagnosed whenever the orbits are circular (or near circular) rather than the usual elliptical orbits.

Therefore, because of this property, the alteration of Mercury's orbit should be thought of as a nielucion (or orbitar precession) of the orbit rather than as advancement within the orbit.

The nielucion of Mercury's orbit is another example where "***Anglewarp Concept***" is in agreement with general relativity (minus the space warp.) This will be discussed further in "***Gravitation Concept.***"

Similar nielucion is to be expected in other planetary orbits. The example of Mercury is extreme due to its nearness to the sun and the increasing effect (angle) of the spinwarp with decreasing distance from the rotating orb.

In the case of Halley's Comet, we should expect unusual spinwarp effects since its orbit is retrograde: The retrograde nielucion of Halley's Comet tends to counter-rotate its orbit. This has the same dynamic effect as a slowdown of the comet's speed. Reduced speed causes inward relocation of the comets orbital path. However, the inward shift mitigates the speed loss.

In other words, inward relocation causes speed gain while speed gain opposes inward relocation. (Please notice the triple-negative prevention. "Losing speed does not oppose inward relocation." This is really the point of the precise wording.)

A Very Large Spinwarp

The rotation of a galaxy is less rigid and thus more complicated, but it is still maintained by spinwarp. Fabrons of the space network would not be expected to rotate with the galaxy.

However, the tremendous mass of an entire galaxy implies a very massive warp field that does rotate (each component independently) with the galaxy. This spinwarp also provides the nielucion that induces orbital revolution of all nearby matter and dark matter. This may cause the total mass of a galaxy to appear to be much greater than it really is.

For example, consider the orbit of Mercury around the sun. If Mercury were less visible, its mass might be mistakenly included in the total mass of the sun due to the sun's orbitar nielucion of Mercury's orbit. Since the entire space network is filled with much dark matter associated with cosmic gas, dust, and elementary particles, a conscious decision must be made as to how much of the dark matter orbiting a rotating galaxy should be considered "part" of the galaxy. The discovery of nielucion brings a weighty perspective to all rotation.

A third option might split the difference, for example, by maintaining a minimal Milky Way galaxy while christening a much larger (and darker) "Greater Milky Way" that would include all (or much) of the "suburbs." (Nielucion will be discussed further in "***Gravitation Concept***".)

Fluid Examples of Anglewarp

Consider the example of water moving through a fire hose. Afterwarp is responsible for the momentum of any fluid traveling in a hose or pipe as it is in a canal or river. Water pressure is generated by pumps or gravity feed at the water source. Because a fire hose holds the water in place, then by Newton's third law of motion, the pressurized water causes a reaction force on the fluid that is equal and opposite to the water pressure on the hose.

When the water in the hose is stationary, the anglewarp imposed by the hose on the water can be thought of as being directed inward perpendicular to the hose and thus perpendicular to the stationary water.

When the nozzle constricting the pressure is released and the water is allowed to flow, motion of the water through the hose generates afterwarp. As the speed of the water flow increases, the anglewarp will begin to slant at an angle toward the direction of flow. We will call this angle that increases with speed the "speed angle."

The component of the speed angle perpendicular to the hose will be called the "stress component of the anglewarp" or simply the "stress warp." The afterwarp is directed parallel to the hose and would thus have no component perpendicular to the hose. This *narrative #8* is validated by our understanding of "water pressure" removing it from the realm of regality.

The component of the speed angle parallel to the direction of flow combined with the afterwarp will be called the "speed warp." The speed warp produces the linear momentum, which will carry the water past the end of the hose.

It should be expected that the reaction on the hose to an increase in the speed warp would be a backward force on the hose. This explains the need for one or more firefighters to prevent the hose from being propelled backward in reaction to the water being propelled forward.

Furthermore, an increase in water speed causes an increase in the angle warp in the direction of the increased speed. This increase in the speed angle results in an increase in the speed warp component and decreases the stress component of the angle warp.

This decrease in the lateral component of the angle warp is generally interpreted as a decrease in the lateral water pressure. This *narrative #9* describes the phenomenon known as "Bernoulli's principle", formulated by Daniel Bernoulli, and explains how it works in terms of its anglewarp removing it from the realm of regality.

Air Foil Example

The anglewarp of the air passing over an airfoil (wing) can be thought of in terms of two components. The component perpendicular to the wing will be called the pressurewarp and the component parallel to the wing will be called the speed warp.

The air passing <u>*over*</u> the wing has increased speed caused by the greater distance over the curved airfoil compared to the distance traveled by air passing under the flatter portion of the airfoil. Therefore, the anglewarp above an airfoil has a reduced downward force compared to the upward force of the anglewarp below it.

Decreasing the downward force on the wing, more than the upward force on the wing is decreased, has a net effect of increasing the lift on the wing. This application of the Bernoulli principle implies that the upward pressure on the airfoil is greater than the downward pressure. This explains the lift on an airfoil.

The decreased pressurewarp of the air passing over the wing could be called a partial vacuum so that airfoil lift could be described as vacuum lift. On the other hand, vacuum relates to matter loss whereas decreased pressurewarp relates to energy loss.

This lift does not work in space because a spacecraft does not travel through its warp, but with it.

The fire hose example generates some interesting questions. Is it possible that the warp field, being energy, is entirely contained within the hose? If so, how does it work? The warp field affects the hose and it controls the motion of water within the hose. However, it is difficult to tell if the motion of the hose is caused indirectly by the motion of the internal water or directly by the internal warp. As the anglewarp maintains the water pressure within the hose, how does the anglewarp on the water affect the hose?

Response

Every firefighter knows that the warp pressure on the hose can be considerable. When the water is first turned on, the vectorwarp response arrives with the water. If a fire hose is not under control when the vectorwarp arrives, it can cause damage, injury, or death.

Gravity Assist

It seems likely that an understanding of how gravity works could shed light on the workings of a "gravity assist." Since gravitation is caused by pressurewarp, then how might pressurewarp affect gravity assist?

It is clear that the gravitation caused by a nearby object is not shielded by any intervening medium. On earth, there is no known "container of weightlessness." There also seems to be no shielding against the effects of any afterwarp. Thus, it seems that the effects of a warp field are not bound by material barriers. Is it possible that pressurewarp holds the key to understanding gravity assist?

Just as pressurewarp causes (angular) precession of the rotating object "carrying" the anglewarp, pressurewarp could also cause "linear precession" of the (linear motion) carrier of the afterwarp. The gravity assist of a space vehicle might satisfy all of these requirements.

The question is, "Would vehicles experience drafting caused by afterwarp if there were no air or water to mask it?" To be more specific, is afterwarp: totally, partially, or not responsible for the gravitational "slingshot" used many times in planetary missions? It seems logical that when a warp field such as gravitation is involved in a so-called "slingshot" effect that this would be a prime example of linear precession. The problem now is distinguishing between gravitation and linear precession, if there is a difference.

It seems logical that better understanding of this slingshot effect might lead to usage that is more efficient. This would be especially noticeable if the effect were called "linear precession". There will be more on this to come. This concept will be labeled narrative #10 removing it from the realm of regality.

Orbitar Momentum

The definition of orbitar momentum is just what it seems. Orbitar momentum of object A is very similar and has a similar narrative to angular momentum of some object B if objects A and B are similar, except that B is rotating (circularly) around its center of gravity (near the center if B and A is revolving (elliptically) around its (gravity linked) central object.

Maximum Anglewarp

Just as maximum afterwarp results in maximum linear speed called "c," similarly, maximum anglewarp results in maximum angular speed that will be called "ω-max." Maximum angular speed of object x located at radius r theoretically occurs when the linear speed at radius r tangent to the circle of rotation is equal to c. This can be calculated by the formula $\omega\text{-max} = c/r$ (radians per second).

Author's Note

"Anglewarp" explains many things such as angular momentum, rotation stability, and precession. The realization that anglewarp explains precession was one of the first discoveries that convinced me that at least some of my budding **Motion Unification** concepts (involving suspected "warp fields") were valid. (This realization occurred while I was an undergraduate student at Baylor University in the fall of 1962.)

Vector Use

The use of mathematical vectors to decode the intricacies of applied physics has been crucial to the "mathematical development of applied physics." The use now of a physical *narrative* to explicate the intricacies of the applied mathematics might be called the "physical epitomizing of applied mathematics." However, in today's lexicon of physics, applied mathematics of the physical world is called "physics." "Pure physics" or "operational physics" cannot be characterized as "theoretical physics" since that name has already been taken to refer to mathematical modeling of physical systems.

The most noticeable difference between theoretical physics and operational physics is the set of tools that are used for the discussion. Since theoretical physics is primarily relational physics, then the primary tool for communication will generally be mathematics. However, the primary tool for explaining how systems operate will generally be a language such as English. This approach is somewhat new to physics. As you know, we are calling it "narrative physics."

Classification of Angular Warp

Angular momentum is classified as a "direct primary action" because rotation of the anglewarp is directly caused and maintained by rotation of the object.

Angular reaction is classified as a "direct secondary action." Linear or rotational motion of one object can directly induce linear or angular reaction in a second object depending on the cause of rotation and how the second object is free to move.

Angular Warp Unification Table 2	**Primary Action**	**Secondary Action**
Direct Cause (Anglewarp)	Angular Momentum	Orbitar Reaction
Indirect Cause (Pressurewarp)	Precession	Nielucion

Angular Precession can simply be called "precession" when the meaning is clear. Precession is classified as an "indirect primary action" because an action on the rotational axis of a spinning object can indirectly induce the object's precession.

Further Classifications

The following paragraph is included to complete table 2. This motion will be discussed later. Note that this table is probably not complete and is in need of further refinement. Trying to fix that in these pages would only introduce new concepts to consider.

Nielucion is classified as an "indirect secondary reaction" because the spinwarp field, of the rotating object "A", indirectly creates an orbital anglewarp on any object "B" that might be in orbit around A. This induces nielucion (orbitar precession) of B's orbit around A.

We have long known that orbiting objects were in gravitational freefall around the center of gravity. We now see that the orbit (in this case, Mercury's orbit) can itself experience rotation about the center and tangent to the central object's equator, depending upon the angular momentum of the rotating object.

Note that Mercury's orbit can have any eccentricity. However, the rotating sun's celestial equator is circular with the sun at the center of the circle.

Thus, while remaining in its eccentric and inclined orbit around the sun, mercury (and consequently its entire orbit) is also being dragged by nielucion around its circular intersection with Mercury's orbital plane. This nielucion is a new concept in this book. Nielucion will be discussed in "***Gravitation Concept***."

3rd Uni-Verse

"The recognition of 'nielucion'
Brings gravitas to revolution."

4. Expansion Concept

Variability of Universal Expansion

In 1929, Edwin Hubble discovered the cosmological redshift and used it to establish the rate at which the universe was expanding. It is difficult to dismiss the supposition that the Big Bang was a giant, primordial explosion. Regardless of the cause, it has been generally assumed that, under the influence of gravitation, the expansion of the universe ***must*** be slowing.

It now seems unlikely that the Big Bang expansion involved a shock wave. This is important since it is unlikely that the expansion of the universe is directly related to any explosive "Big Bang."

Like the "Big Bang," the "Challenger disaster" is often referred to as an explosion. Since there was no shock wave, the NASA review team knew that the shuttle did not explode. Instead, it was torn apart by its own slipstream when forced by structural failure from its intended trajectory. The fuel from its ruptured tanks, some of which briefly burned, was released into the thin atmosphere giving the appearance on home TVs of an explosion.

An explosion was actually unlikely due to insufficient heat, and more importantly, insufficient atmospheric oxygen at the high altitude the shuttle had already attained.

The Epoch of Inflation

What we still call the "Big Bang" was a brief instant of extreme expansion known as the "epoch of inflation." Still a mystery (until now), the "epoch of inflation" was hypothesized in 1980 by American physicist Alan Guth.

This expansion may have been even faster than light speed. We are told by cosmologists that the universe underwent an unfathomable expansion by a factor of at least 10^{50} times (regardless of the unit being used for the measurement, just affix 50 more zeroes to the end of the numeral representing its former size) in just 10^{-32} seconds.

In unusual circumstances, it is much easier to brush aside the ultimate speed limit c if you do not know why it is generally obeyed. This will be revealed soon.

It is noteworthy that the universe was so small initially that (assuming exponential expansion even then) nearly half of the expansion time was used just making the universe large enough to hold one small person. Talk about cramped quarters. It was a good thing there were no people yet. This was (by some estimates) roughly 13.8 billion years ago.

Prior to this publication, neither the source nor the process of this expansion has ever been adequately explained. (The concept of a "false vacuum" seems acceptable as a framework for computations, but this fosters little understanding and provides no narrative. Just because it works and even affords a few answers is no substitute for a narrative.)

The concept of "vacuum energy", even neglecting an enigmatic "vacuum catastrophe," has also not yet been physically explained (even by the Top Quark or any of his/her "Lesser Quarks"). The regal science of inflation is mindful of the concept of precession before precession was finally explained in "***Angular Momentum Concept***" above. The epoch of inflation will be explained now.

Hereafter in this book, the term "vacuum," unless quantified, will generally refer to a partial mass-vacuum of the fabron-filled space network.

The Big Blink

The "really fast" incident called the "Big Bang" could now (but, with respect to tradition, will not) be called the "Big Blink." In 10^{-32} seconds, it was finished and "done with" incredibly fast, even compared to regular explosions.

The Big Bang occurred at an unusual time. Enormous heat and expansion would be expected in a major explosion.

However, the phenomenal heat of the Big Bang, rather than being generated by an explosion, was likely building up long before its seemingly uncontrolled release. The heat's "residue" must have then progressively diminished with universal expansion.

The conditions seemed sufficiently similar to a "Big Bang" to motivate its naming by the incredulous Sir Fred Hoyle on March 28, 1949.

In keeping with the simplest concept pledge, let us consider that if any object or action physically existed before time began, even for a twinkle of God's eye, there was also probably potential energy, potential dark energy, or perhaps both if there was a difference at that time. However, without specifics this is not helpful.

Apparently, the conditions somehow became right for the formation of the space network out of the fabrons (dark energy) of space.

(Even from the current scientific point of view, it is unknown what the conditions were and whether they were internal or external. It is also unknown where the energy came from, how, or when.) Some of this will be discussed later.

Due to the presumed 1) small space, 2) high temperature and 3) homogeneity, it is probable that the space network formed everywhere concomitantly in an incredible expansion by a factor of about 10^{50} in a process called the epoch of inflation.

Of course, "everywhere" was initially a very tiny smaller-than-a-speck "singularity." Particle physicists tell us there is no limit to the number of photons that can simultaneously occupy the same tiny singularity.

This insight does not seem so outrageous when we realize that we are talking about energy such as photons or fabrons (if either of these is truly massless) and begin thinking of it mathematically. [Just the same, even a "finite" singularity of energy (if "full" enough) might do wonders for our occasional energy crises.]

It is somewhat reassuring to know that the source of the heat is (almost) irrelevant to the remainder of the creation process. The very high concentration of radiant energy that was released would automatically have had an extremely short wavelength. Even without ignition, this is considered an extremely high temperature comparable to an unimaginably monstrous explosion otherwise.

Later in the "***Advanced Speculation***" section we will ponder the possibility that the "stacked" energy within a singularity might be polarized, coherent, or both and the possible consequences of these prospects. We will call this condition, "singularized."

Consider this question, "How might 'polarization,' or even 'spatial or temporal coherence' affect the initial expansion of inflation?" Here are some of the most common assumptions supplied by cosmologists concerning the conditions related to inflation.

1. At the point billions of years ago where our understanding of inflation begins, the size of the universe was probably below 10^{-25} m. This is less than one trillionth of the size of an atom. We will call this the "micro-verse."

2. The temperature of the micro-verse has been projected to have been in excess of 10^{28} K. This is about 10^{21} times hotter than the sun and science cannot tell us where the heat came from.

3. The epochs called the "grand unification epoch" from times 10^{-43} sec. to 10^{-36} sec. to the "electroweak epoch" from times 10^{-36} sec. to 10^{-32} sec. after the beginning of time saw the cooling of the universe to the level in which all of the Grand Unified Forces separated into the individual forces as they are today.

4. The epoch of inflation supposedly began at 10^{-32} sec. after the beginning of time but lasted for an unknown tiny fraction of a second.

5. In a mathematically poetic sense, this 10^{-32} sec. is one hundredth of a thousandth of a millionth of a billionth of a trillionth of a second. For perspective, just remember that one trillionth of a second is exactly one millionth of a

millionth of a second and one millionth of second is exactly one thousandth of a thousandth of a second, etc. To complete the break-down, remember that one thousandth of a second is exactly one tenth of a tenth of a tenth of a second.

6. It now seems likely that the epoch of inflation did not "follow" the Big Bang, but that it more fittingly fits the description of the Big Bang itself or at least included that. It is assumed that the "big picture" universe was and still is homogeneous to this day.

Our first point of elucidation will be at step #2. Clarity here requires a distinction between heat and temperature. *Heat* is a measure of the ***quantity*** of energy, whereas *temperature* is a measure of the ***intensity*** of the energy.

A matchstick and a campfire, both of the same type of wood, will burn at roughly the same *temperature*. The campfire, however, will produce many times more *heat* (energy).

A temperature of 10^{28} K is an extremely high temperature. However, due to the very tiny size, it is not as much heat as it may seem, at first. This will become an important factor in our understanding.

The wording of #3 implies that the beginning of time is being thought of as the initial production of energy. The narrative does not seem to support this and it will not be assumed by **MU**.

Instead, **MU** will associate the expansion at the beginning of time with the expansion during the epoch of inflation. This was partially explained above.

The next point of contention is #6. Because of its extremely small size, the micro-verse was most likely homogeneous initially. The distribution of energy, throughout the version of space assumed by **MU**, was likely accomplished by the space network.

There is neither narrative nor evidence that homogeneity was even possible in the initial expansion before the existence of the space network.

Our understanding of the early cosmos requires an understanding of the conditions at that time. In "***Advanced Speculation***," we will address some of the possible conditions before and at the beginning of time. It is tempting to say that the individual particles of the space network could not have been connected, coordinated, or even related since "there was not time" to establish those relations.

This tells us nothing since "there was not time" to become so huge either. Yet, that apparently happened. This "flexible network" concept is simple and will be assumed, but it only seems to work if we stipulate a self-adjusting energy field associated with each fabron. It is difficult to break this process down into fathomable steps that each took tinier fractions of these minuscule fractions of the first second of "time."

Stability of Expansion

Before the narrative of the space network, there was no *narrative* for the fixed density (or "stability") of galaxies traveling through space. The general attitude, with notable exceptions, has been that "they are-what-they-are and will generally continue to be-what-they-were" with modest, natural mutability.

The pledge of "strongest heart" would seem to require 1) as the universe expands, that the density of the fabrons remains constant and 2) the geometry of the universe remain flat everywhere.

Requirement 2 will be explained in "***Cosmic Flatness Concept***." Requirement 1 will be partially explained now.

Motron Stability Conjecture

We will assume that the space network is composed of motrons with fixed "density." This condition is being referred to as motron "stability." The word "density" is being appropriated here to refer to measurements such as motrons per cubic micrometer.

In other words, during inflation, motrons established a stable universe with uniform specific density D. There appear to be at least three possible alternatives for the current value of D.

Option 1 is the circumstance of the arbitrary value of D at the moment when extreme inflation ended.

Option 2 is the idea that the universe requires stability, but the specific value is an average value likely related to the naturally occurring, Planck length, named for German theoretical physicist Max Karl Ernst Ludwig Planck. This will be called "density averaging." The Planck length is **D** $\approx 2.36872852 \times 10^{86}$ motrons per cubic micron (mm^3). The process by which the universe is able to maintain nearly constant density D for the entire space network will be called "D-pushing."

Option 3 suggests that a possible value for D might be expressed using the mass equivalency formula $E = mc^2$ to compute the equivalent density as 6.91×10^{-27} kg/m^3.

However, the actual value of D is immaterial to the logic of our study; its value only matters physically.

The Euclidean Network

Until specific significance of the current value of D becomes known, "***Expansion Concept***" will simply assume option, #2. The specific average value of D will be discussed in "**Advanced Speculation**." It must also be mentioned that the outward pressure of D-pushing based on density averaging is also probably at least partially responsible for cosmic acceleration. Dynamic flatness and its relevance to this point will be explained in "***Cosmic Flatness Concept***." Please see D-pushing in "***Advanced Speculation***."

For now, it should be sufficient to point out that universal stability implies Euclidean-like flatness for the energy of the space network. It should not stretch credulity to suggest that the space network of energy should have the same "geometry" as material space. It is, in fact, difficult to imagine otherwise, "space warp" notwithstanding.

Photon Pressure

The *narrative* of universal stability brought by uniform (fixed) density will partially explain why the parochial speed of light must be a constant, which we call "c."

Since innumerable photons are constantly in motion in all directions throughout the universe, and geometry is probably Euclidean everywhere, any change in density D would seem to require an abrupt c-change throughout the universe. (Pardon the pun.)

Lacking the ability to change the value of c instantly throughout the universe might constrain the universe (as we know it) to constant density. Regardless of the mechanics, this constant density is the essence of the "Motron Stability Conjecture."

This conjecture cannot be physically verified yet, but the ramifications of this one assumption lead directly to new understanding and possible answers to many long-held questions. This is the essence of a successful inductive proof.

It is interesting to note that a change in motron density would change the speed of light passing through space. However, the converse of this conditional statement is of little interest unless there is some other factor besides motron density that could change "parochial" (local reference frame) light speed through a vacuum. This seems very unlikely if all photons are alike, all warpons are alike, all fabrons function in the same way and at the same speed, and the average value of D is constant.

Transition from Puddling to Pooling

These two terms, as understood in **MU**, will refer to opposite circumstances. In both cases, the contents of a set are unchanged. Only the distribution is altered. With *puddling*, initially, the contents of a new warp field are concentrated in *many* locations (puddles). In the transition to pooling, the over-all density remains the same. The contents (the pools) become *fewer*, *larger* and *more separated.*

Singularity Reduction

It is likely that warpons exist as individuals rather than in bundles (singularities) as fabrons likely do. Fabrons are capable of multiplying by subdividing. This process will be called "singularity reduction." The reduction (or breakdown) of a singularity of fabrons into individual particles will be called a "total singularity reduction." Both types of reduction will be pondered in "***Advanced Speculation***."

A less drastic version of singularity reduction may also be attainable. It may be possible for a singularity to "dole" a single fabron at a time in a process we will call a "singularity dole." This might be used to explain how gravitation, alone, is able to prevent galaxies from expanding while the space network is expanding around them. This, too, will be pondered in "***Advanced Speculation***."

"Clumpability" is a serious technicality within the Motron Stability Conjecture. The clumping problem will be addressed with "pooling" in "***Advanced Speculation***." Henceforth, we will assume (justification to follow) that fabrons exist in singularities, but warpons are now (since matterfacation) always singletons. This *narrative* #11 explains "Universal Stability" for the first time, removing it from the realm of regality. The process of singularity reduction will also be discussed in "***Advanced Speculation***."

It must be pointed out that constant density does not imply constant, linear motron separation. Motron separation will generally be called "span." In fact, even with constant density, in any two-dimensional subspace it is impossible for linear motron span to be uniform in any distribution.

Simplified and Magnified

Space Network

(Plane Fractal Model of "***VAMOS***")

Figure 1

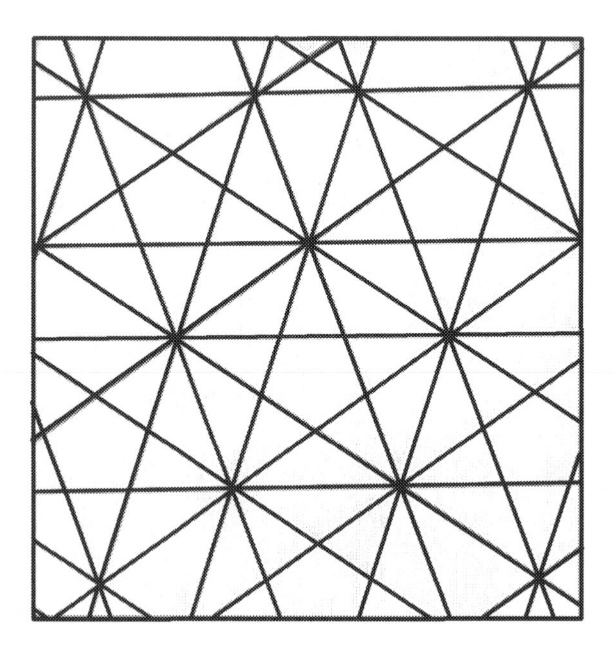

<u>Va</u>riable <u>Mo</u>tron <u>S</u>pan

Imagine any line you might draw across the simplified space model in Figure 1. Notice that your line passes through or near many vertices. Note also that the distances between adjacent vertex encounters is a variable. This phenomenon will be called, <u>va</u>riable <u>mo</u>tron <u>s</u>pan or "<u>vamos</u>" for short. This is even more obvious in a somewhat less-magnified 3-space (not shown).

Average fixed density of fabrons (or stability) suggests a new unit for measuring speed of passage through the space network.

Light wave frequency cannot be used since speed is independent of wavelength, λ, and of frequency, f, (except that, with appropriate units, λ and f are reciprocals of each other so that $c = \lambda \cdot f$). One possible name for the new unit might be "cogs per second" (cog/s). Since this is a new unit of measure, it can be arbitrarily set as 1 cog/s = 299 792.458 km/s or (about) 3×10^5 km/s. The word "cog" is taken from the name for a gear tooth. (These are numerous in mechanical time pieces.)

The precise value required would result from taking the value of c as exactly 1000 cog/s or ($c = 1 \times 10^3$ cog/s = 1 kcg/s). For perspective, we can estimate that 1 cog/s which is approximately 300 thousand kilometers per second is also roughly 3 billion-billion motrons per second. The precise value is irrelevant to this discussion apart from the simplification it allows.

The concept of vamos produces an unexpected corroboration. As just mentioned, a particle traveling a very short distance at a constant rate in motrons (vertices on the network) per second (as is normal) would not accurately make constant progress in terms of distance per second. This means that the exact microscopic location (vertex) of any moving particle could not be calculated accurately at any exact time unless the particle was large enough or traveled far enough to make the argument moot. This *narrative* #12 seems to explain the "Heisenberg Uncertainty Principle" which was named for German Physicist Werner Heisenberg. The Heisenberg Uncertainty Principle will be shortened to "HUP."

Confirmation of this *narrative* for HUP might establish the "Motron Stability Conjecture." This, in turn, would establish "***Expansion***

Concept." The ramifications beyond even "***Expansion Concept***" will be revealed as they are engaged.

This *narrative* dissolves the negative connotation from the uncertainty of "HUP" removing it from the realm of regality. HUP could be restated, "Due to the variable motron span (vamos) there will always be a minimum precision unit of the Planck length $\ell_P = 1.61619926 \times 10^{-35}$ m in the distance and direction traveled by any particle moving through the space network." The margin of error could be called the "radius of confidence" in any distance measurement; its value would thus be ½ ℓ_P.

There is one additional possible ramification of this tiny inner radius of confidence. These tiny "gaps" in the space network might allow small particles with spin to rotate inward through the gap causing the appearance of an orientation reversal (inside-out) when seen from the opposite side of its invisible "local surface." Alternatively, this might be imagined as a rotation through a thin 4th dimension. This *narrative* may diminish the expectation of an "actual" fourth dimension.

Note that the space network is not just a support system around space or even a support system within space. The space network is more than just the fabric of space. The space network is the "space continuum." This should help us understand how the space network is responsible for the "flatness" of the universe.

Speed of Transmission Through the Space Network

"***Expansion Concept***" will assume that, to this day, the primary interaction between fabrons of the space network is their temporary connections to the warpons of any common massive object. In other words, if two fabrons are presently providing momentum to the same object, then this is also a momentary bond between the two fabrons.

As the universe expands, motion of objects along with their warpons continues to work in the same way and at the same speed.

Travel in the expanded universe takes longer, however, due to the increased distances. This partially illustrates how parochial light speed c is unaffected by any motions of the source or of the receiver. There will be more on this to come.

Note that the expression "light speed" will refer to the mean *parochial* speed of light that is known to be constant c. This was seen to be constant in the Michelson Morley Experiment and many other similar experiments. This result will be called "parochial constancy."

The unmodified expression "speed of light," however, will include alien speed that, in "forced relativity," is assumed (forced to be, but never proven constant when making computations) in addition to parochial speed being constant. The difference will be discussed in "***Relativity Concept***."

If photons and other objects travel through space (with speeds measured in kcg/s), then average speed for even relatively short distances would depend only on motron stability. This corroborates the motron stability conjecture.

Imports of "*Expansion Concept*"

There is no particular significance to the specific fractal model presented previously in figure 1. The concept would be the same regardless of the actual pattern of the space network. The theory works even if the tiny pattern is either nonexistent or constantly shifting. However, we must at least consider the possibility that the basic framework of the space network could be some type of fractal based upon its internal singularity microstructure, if any. This could have ramifications down the line since much of nature seems "fractalian."

To be more precise, a pattern is not even necessary for vamos. The only requirement for vamos is the presence of an average-stable space network. It is probably best if the network is thought of as points that are not connected by lines. The lines only help see the points and the pattern better.

With lines, it would be logical to think of a *spherical* universe as being Riemannian. Without lines, it is much easier to think of an *expanding* universe as exactly Euclidian. (More on this in the next chapter.)

Vamos seems the simplest explanation for HUP. In return, HUP seems another strong verification for the presence of a space network. HUP still laments what we cannot know, whereas vamos now proclaims what this reveals to us.

Regardless of the network pattern, whatever "pattern" might exist is likely distorted temporarily by "effectively" stretching and straightening as photons pass through. This phenomenon will continue to be referred to as "photon pressure."

The linear result of photon pressure will be referred to as "linear straightening." The 2-D result of linear straightening will be referred to as "plane flattening." The 3-D result of plane flattening will be called "dynamically flat", or "d-flat" for short. D-flatness will be discussed soon.

Please note that the <u>mean</u> margin of error must be constant in order that light speed c also is constant. Notice also that the HUP error in absolute c is very tiny compared to the very large mean parochial value for c. Mean c can thus be "considered constant" to any precision we are able to achieve.

"***Cosmic Flatness Concept***" (next) explains the circumstances that require the rate of expansion of the universe to be increasing. This also explains the "flatness" of the universe. More importantly, it explains how this is accomplished, how expansion, along with photon pressure, straightening and flattening, and even the parochial constant c, are related to flatness and the need for flatness.

4th Uni-Verse

"Men will often judge us
By where we seek our gains;
Expansion of our homes,
Our billfolds, belts or brains."

5. Hyperspace Concept

The Parameters of Space

We will assume that space consists of 3 geometric dimensions that are always finite. This will be discussed in "The Cosmological Principle" below. Any desire for a time dimension to accommodate relativity theory and the codependent cosmological principle will be discussed in "***Relativity Concept***" later.

The value of any parochial time variable will always be current and universally consistent. This makes it a universal parameter. The passage of time will be assumed a universal constant.

To support their mathematical understanding of a space warp, Albert Einstein and others felt (rather arbitrarily) compelled to attach a time dimension to physical space and call space the "space-time continuum." This will be discussed in detail below.

Although rather complex, the concept seems to work mathematically. There is currently no physical *narrative* to support its physical necessity, however. This will also be discussed soon.

Meanwhile, those who would like to "see" a time dimension attached to the space continuum should use their imagination to visualize it. This is how it must be generally done in real-life when time is a factor in a particular function.

Time is not a geometrical dimension, however. Geometrical dimensions are sometimes fixed but allowed to change as needed, whereas time is subtly advancing in reality even when temporarily imagined as fixed. For the purposes of a time related graph, a 2-D, 3-D, or even a 4-D time-graph with one or more geometric dimensions are possible.

Moreover, time is not physical like space. For any point in space, time (as it exists) is always current. Barring shackles, it is always possible to move around in space. However, nothing can be shackled to time and yet it is impossible to move around in time, especially when fixed in 3-space.

Time never stands still, but everything material must stand <u>currently</u> still in time. In other words, the three dimensions of geometry are independent of each other, but dependent on time. For any specific value of time, motion is impossible. One independent variable and 3 that are dependent is not "hyperspace" or even physical "space," it is most likely a (mathematical) "hypergraph," even if that graph is called "space-time."

A hypergraph of 4-space requires each point of space to have 4 dimensions that are sometimes called "hyperedges," and points of 4-space are often called "nodes" or "vertices."

<u>Physical Space</u>

There is one additional consequent of hyperspace. Since time is not a geometrical dimension, then hyperspace is a <u>mathematical</u> model of so-called "space-time." Since this model is more mathematical (mental) than physical, then it should be used when needed to compute a mathematical result or communicate a mathematical concept. However, it cannot be promoted based on physical reality.

Early Greek mathematicians seemed to understand the difference between (mental) geometric (concept) constructions and physical models constructed in some media.

They demonstrated this clearly with their physical models of concept constructions. [E.g., the Euclidean concept compass cannot retain its radius (as opposed to a physical compass) when lifted from the paper.] However, the concept compass is superior since mental arcs can be considered exact while (physical) compass arcs, like (physical) straightedge segments and measurements, are always approximate.

For convenience, integers and exact measurements can be rounded to any desired precision. Whereas, (physical) measurements must be rounded to the measured precision or more.

The Cosmological Principle

Most physicists assume the twin principles of homogeneity and isotropy. These concepts sometimes encourage the use of hyperspace. This is not difficult for linear or planar motion since the unused dimensions need not be visible.

It does complicate otherwise simple motions such as cosmic expansion or acceleration. For these reasons and others to be discussed later, **MU** will only use 4-space when it cannot be easily avoided. It should be obvious that in 4-space any of the unused dimensions can be ignored. This includes time.

The concept of d-flatness in 3-D is simpler to explain than trying to explain flatness in 4-D. Many have tried to explain 4-D; this will be discussed shortly. The ability to explain space-time using hyperspace is supposed to imply that space-time can be considered flat.

When all three geometric dimensions are used along with a fourth dimension of time, it becomes difficult to distinguish between "what is" and "what is to be." This is unfortunate because this is exactly "what the 4th dimension is to be for (sic)."

In "***Relativity Concept***", we will show how 4-D space-time can generally be avoided.

This seems a good place to affirm the other dynamic properties of space. No (practical) *narrative* permits a finite universe to become infinite in a finite amount of time.

Like flatness, the other three dynamic properties all require a fourth dimension of time. The assumptions of homogeneity and isotropy both require an infinite universe to prevent space from having boundaries.

However, all three of these properties are at least possible in a dynamic 3-space. In a dynamic 3-space, boundaries are not a factor since boundaries can never be reached when traveling with time in such a space.

Disarming the Space-Time Bomb

Even though space-time is required for the current version of the cosmological principle, this is not the only reason space-time was introduced into physics.

The cosmological principle is now based on the conjecture of "space-time." Much evidence suggests that (at least in some quarters) space-time first became a requirement for the "equivalence principle" as constituted within the desired framework of general relativity.

One important distinction between the frameworks of gravitation and of momentum is that gravitation provides a massive object with acceleration, whereas simple momentum generally requires only speed. The difference between acceleration and speed is a second time component.

Before the space network was discovered by its effect on motion (as revealed in this book), space warp stood in for it. However, space warp does not provide the time component that is necessary for any equivalence between gravitation and motion. This seems to have been one of the original motives for affixing time as a fourth dimension of space, thus creating "the space-time continuum." This equivalence will be called "gravitation equivalence."

This method of introducing a second time component into the equation obscures the fact that centripetal acceleration requires mass just as linear acceleration does. This implies that space warp (which has no narrative) can deflect (bend) motion in space just like gravitation does. (The narrative for gravitation will be introduced soon.)

Since aether was "debunked" and the space network did not yet exist, the invention of space-time was the simplest choice that was then available. This was a near-miss.

However, miss the mark it did. This one "ricochet" produced not only space-time, but it eventually led to a misunderstanding of gravitation, mass, equivalent mass, alien c, special (forced) relativity, $E = mc^2$, and probably other concepts as well.

This finally divulges why so many concepts of physics cannot yet be explained. Most of these concepts will (especially in this document) survive with only moderate modification. Some will not. These will be discussed in the remainder of this book.

One other unfortunate implication of a time dimension that mimics distance is the false implication of forward and backward (physical) time travel. This is a great boon to science fiction writers. This "gift" may be even more desirable than the sci-fi dream of a fourth geometric dimension and a "space warp" to get there.

They should not lose hope, however. The multiple dimensions of "string theory" still live in the hearts and minds of many advocates. Unfortunately, no *narrative* has been provided for any of these concepts. (*Narratives* seem metaphoric among vibrating strings.)

In "***Gravitation Concept***," **MU** will provide the *narrative* for a "<u>revised and expanded</u> equivalence principle" beyond the confines of space-warp and general relativity. It must be noted that until now, the discipline of physics has never been closer to a true understanding of gravitation than when Albert Einstein introduced the equivalence principle a century ago in 1907. This concept actually grew from discussions initiated by Italian astronomer Galileo Galilei. This notion will be explored in "***Gravitation Concept***."

<u>So What If the Universe Isn't Flat?</u>

Flatness will be discussed in "**Expansion Concept**" below. To some it may seem sufficient for the universe to be so large that it "<u>seems flat</u>." If some think the universe "<u>seems flat</u>," they must be thinking of the *<u>surface</u>* of the universe. Even the earth often seems flat when standing on its surface. However, we (most of us) live on the surface of the earth and so we (most of us) ***know*** that it is not flat.

The problem with thinking that the universe seems flat is that nothing is on the <u>surface of the universe</u>.

All visible objects, including all galaxies, are "inside." Furthermore, many astronomers consider the universe infinite; therefore, for them there would be no "natural surface." Being infinite, of course, does not make it flat, just less visual because of its invisible time dimension.

No, looking at a clock does not make time visible. Clocks, watches and even sundials only provide the ability to uniformly measure the ***assumed uniform**** passage of time in the units determined by the instrument maker as construed by the user.

* *The asys of "time dilation" will be critically explored in* **Relativity Concept**.

One problem with this point of view is that if we suppose that there is no surface, then we tend to mentally create a surface (upon which we can stand) with a low probability that it will be co-temporal (all existing at the same time).

Identifying co-temporal surfaces is actually easier on models than it ever could be in real life. Even if we could travel through space-time, how would we see (into or through) the time dimension or control our travel through it?

It seems clear that we can only travel through 3-D space. Allowing time to pass (Who can do otherwise?) does not constitute travel "through" time, but "with" time.

If we ever need to create a surface to stand on, then why not just <u>create it to be flat</u>. Away with that annoying geometry, full speed ahead!

Note that thinking "the universe is so large that it seems flat," might be appropriate terminology within stage #4 of "Static Flatness" within "***Cosmic Flatness Concept***" (next) since it would be in stage #4 that we could consider conditions on the future-surface of hyperspace. At this point in our thinking, we must consider the importance of any unsuspected curvature of our supposed flat surface.

To others it may seem sufficient for an arbitrary co-temporal section (shell) of the universe to be so large that it seems flat. Saying it is arbitrarily true would imply it is true for all co-temporal sections. This would betray a faulty "visualization" of hyperspace.

Don't Judge a Book's Flatness By Its Cover Size

It should be noted that both versions of non-Euclidean geometry can be infinite and neither is flat. However, both "seem" flat on the scale of the earth, the scale of the solar system and far beyond, even to distant galaxies as seen through great telescopes. This should warn us that judging flatness on how it "seems" is dangerous.

We must remember that a co-temporal subspace T of space S is generally one dimension "smaller" than S. When the passage of time is considered constant within any space then this removes time as a variable and thus as a dimension. Therefore, a co-temporal subspace of space-time could have 3 geometric dimensions (or less) and no time dimension.

So, exactly how should we visualize the interior of a 4-D universe? To see the problem, think of the universe as a "cosmic onion." A co-temporal section of this onion model should not be what you get by slicing it. Doing that would present a problem. (No, we are not talking about onion-tears here.) A section-into-layers should be obtained by peeling it. (Just wear a breathing mask if you need to.)

This would allow all points of each layer to be co-temporal. Each layer will have a particular time component. The common thickness of all layers will be called "the onion scale." (This should not be confused with "union scale.")

Since a temporal dimension is not the same as a geometrical dimension, then our onion model of hyperspace is not abstract. Specifically, the dimensions are not arbitrary and are thus not interchangeable. We will use "contraction inward," called "backward," for the negative time dimension and "expansion outward," called "forward," for the positive time dimension of 4-D.

If the onion is large enough, the first layer might be flat enough to fashion paper out of … onionskin paper, of course. However, the curvature will become noticeable in due course and it will eventually become extreme.

Some might dismiss this problem by noting that when peeling the cosmic onion you are going inward (backward) to a time where the universe was not yet flat.

If this were true then the universe must also have been flat back then if it were flat now, because at any point in time it can be represented by any size onion.

Trying to deal with the problem in this way misses the point. 4-D is more complex than just deciding, "How flat is flat enough?" We must also ask, "How do we define flatness in 4-D?" In addition, "What role does time play in our definition of flatness?"

We must also decide the meaning of infinite. Does every geometric dimension of an infinite universe need to be infinite, or just the time dimension? (The difficulty of imagining space-time with finite 3-D space and infinite time illustrates some of the complexity of space-time as a tool.)

Peeling the cosmic onion actually presents at least two more problems. Most of the universe would not be flat, and very little of it would be homogeneous or isotropic. The onion layers might individually be homogeneous and the very center of the universe might be isotropic, however.

Once you understand the problem, then you can see a (more complex) similar problem when slicing the onion instead of peeling it.

To avoid some of the problems related to understanding flatness, it may only be necessary to confine our attention to a "small" area of the universe such as a particular planet. Generally, curvature of space may only be noticeable on a very large scale. Even choosing a particular planet could still be a hyperspace unless we choose a particular planet at a particular time.

Wait, we are not done with this concept yet. Most people get confused when we add the 4th dimension of time to a geometric 3-space. The cosmic onion is supposed to be a model of 4-D space-time. The problem we have is the lack of an actual 4th dimension in real 3-space where we can place our model's 4th dimension.

No matter how you try to model a 4-space, your model will always have one or more dimensions of missing data or (equivalently) missing or distorted data in multiple dimensions. A 2-D drawing or photograph of a 3-D model of a 4-D space always has two missing dimensions of data (or the equivalent).

However, even if one dimension is missing out of 4, we still have the other 3. We must make them count for something even if they cannot accommodate everything.

Numerous clever techniques help some, but that discussion is beyond the scope of this book. At this point, it seems obvious that we should try to find a visualization that is 3-D if possible.

A Problem with the Cosmic Onion

In accordance with the "inward/outward" dimension discussed above, we understand that we can let each layer of the onion represents a different (arbitrary) age of 4-D hyperspace. By making the onion scale equal to 1000 years, we can call each layer a millennium. The layers show greater curvature as we work our way back through the millennia toward the beginning of time at the center of a smaller (curved) universe. If we reverse the process and work our way back-to-the-future, we witness "space" getting "flatter."

Q: Since infinite expansion implies that the size of the universe is approaching infinity, does this imply that the universe is approaching flatness?

A: Nice try! However, the onion layers showed greater curvature as we went back in time for two reasons. First, our human size remained the same. This objection is reasonable since humans (probably) do not expand or shrink as the universe expands or shrinks.

The second reason however is more problematic. As the model universe contracts, we remained <u>outside</u> looking at the <u>surface</u> of the dwindling sphere. Let us examine the missing data.

Solution to the Cosmic Onion Problem

Each layer of the onion only shows the outer 2-D surface of each 3-D millennium since time is different from distance, but fixed for each millennium. To understand this, just look at the top millennium.

Can you see the entire 3-D universe or just the (curved) 2-D spherical surface of it? If the surface does not seem two-dimensional, then flatten it out if you have to. If the surface were 3-D you could still "squash" it (like a quarter section from a solid rubber ball), but you would not be able to "flatten" it (as you could if it were a quarter section of a tennis ball).

Now go back one millennium by peeling off the top layer. What do you see? Do you still only see the surface of the previous millennium? As you peel the onion, what happens to the remainder of each millennium?

Because of the missing fourth dimension in 3-space, each new millennium hides most of the previous millennium.

We can probably communicate this concept better by using an analogy from 3-D. Consider a 1 dm^3 stack (where dm^3 is the abbreviation for "cubic decimeter") of paper. (This might be like a stack of square note paper.) Except for the 1 dm^2 plane ("square decimeter" sheet) on top, we can only see the edge of every other plane below the top one because the rest of each plane is hidden by the planes above. By un-stacking the planes (sheets of paper), we become able to see the entire square region of each plane.

Q: How could we see an entire millennium of 4-D space/time?

A1: Instead of a 4-D onion, we could imagine a 4-D hypersphere to be transparent and all interior millennia to be transparent. This model, though impossible to construct due to its 4 dimensions, has some mental advantages; however, for 3-D humans it only serves to accentuate the <u>surface</u> of each millennium.

A2: We might, instead, consider a <u>dynamic solution</u> in which we use a different sphere for each millennium (getting larger for each passing millennium by using a consistent "onion" scale) to represent the expanding universe with passing time. If we use the same scale as the onion that was peeled earlier, then the surface of each millennium should be similar in size and shape to each corresponding layer of "skin" that was peeled off the onion.

Each 3-D sphere looks like different sized versions of a 4-D onion except that, in the 3-D situation, each solid sphere represents an entire concomitant (co-temporal) universe.

It is important to realize that none of the spheres has any "overlap" space in common since each sphere represents a different millennium. Apart from the time dimension, however, they do overlap in physical space. This identifies just part of the difficulty.

Q: How is the dynamic solution better than the 4-D hypersphere?

A: Since each solid, 3-D sphere represents the entire universe at the same time, it is now easy to see that nothing travels on the surface. It is also easy to see that the flatness or curvature of the surface is no indicator of the flatness or curvature within that millennium. We can also see that there is nowhere to insert or imagine an entire (infinite) straight (Euclidean) line.

Furthermore, no matter how large the universe gets we will be no closer to being able to fit an infinite line into a finite sphere. It is thus clear that <u>a finite, static universe cannot be considered flat</u>.

Some may imagine that a great circle of a sphere might represent an infinite line. This might be true in a Riemannian model, named for Bernhard Riemann, but we are talking about Euclidean (flat) geometry here. Also, notice that, as in Riemannian geometry, there are no parallel lines in any 3-D geometry without at least one infinite dimension.

Riemannian <u>plane</u> geometry can be modeled with hollow spheres where lines would be represented by great circles. Parallel lines could therefore be represented by non-intersecting great circles on the surface of a sphere <u>if that were possible</u>. It is <u>not</u> possible because parallel lines do not exist in Riemannian geometry. Let us return to Euclidean geometry.

Q: What data is missing in the dynamic solution?

A: Now, with separate spheres for different millennia, it is difficult to tell exactly how each millennium is related to the next and to the previous millennium. For example, each millennium should still have look-back time. Where is it? Remember that each solid sphere (including its interior) represents the entire universe at the same time.

Q: Does this prove that the universe is <u>not</u> getting flatter?

A: This does not even prove that the universe does not "seem" to be getting flatter. However, the latter might be accomplished with a formal proof that no matter how flat the curvature seems, other parts (closer to the center of the sphere) will be less flat (by a predetermined amount) at the same time. Thus, the 3-D universe is not even homogeneous. The detailed proof of this would be tedious due to the need to quantify "flatness."

Q: What can we learn from these models?

A: We more easily see that the contents of each millennium are inside, not on the surface. We also see that in real time the universe is not homogeneous, isotropic, infinite, or flat.

It is not too difficult to imagine that the universe could be considered infinite if it were in the process of expanding for an infinite amount of time. It is very easy to understand that, while expanding, the universe would always be instantaneously finite.

Most importantly, we see from these models and from the 2-D "Transflat Model" (below), that <u>flatness can be addressed without resorting to a (mentally lavish) fourth dimension</u>.

The establishment of a 4-D space-time continuum is much too pretentious and elitist. It tends to discourage and alienate too many great minds from the formal discussion of cosmic flatness.

Lest the fleet of mind become condescending of the less fleet, let us consider the likelihood that time may cause fleetness to become fleeting. (As a university professor of physics, mathematics, and statistics for 47 years, the author is very familiar with the difficulty of communicating advanced physical and mathematical concepts even to the fleet. Note: Being a professor of the fleet is totally different from the responsibilities of being a "Commander of the Fleet" (such as a Fleet Commander in the Royal Navy.)

Therefore, the “flatness” of the universe is not achieved by having become larger. In the limiting process, the limit, as radius r approaches infinity, of any cross-section of a sphere is a plane. There are two problems with this analogy.

The first problem with this is that “becoming large” is not the same thing as “approaching infinity.” The difference is embodied in the meaning of the word “large.”

The second perception problem in the 3-D “extrapolation” to 4-D involves a 4-D sphere (called a hypersphere). A cross-section of a hypersphere is not a plane but possibly a solid 3-D sphere or more likely a solid figure with the possibility of single-axis rotation, projection, or reflection symmetry (depending on where and how the section is made).

<u>Removing</u> an entire dimension (even time) requires choosing a particular (fixed) value for that dimension unless the removed dimension is independent. In that case, no value is necessary.

If one of the dimensions is time, then it can be removed by setting its value to, “current” or “auto.” The time dimension could also be run from any starting point and at any speed. However, none of these scenarios is 4-D, but three dimensions w/timer. We will call this a “dynamic three-space.”

<u>Summary</u>

To summarize, it is “time,” the process of “becoming” larger that provides “flatness.” To remain flat the universe could continue expanding exponentially, which means precisely at its current rate (with respect to its size). When any expanding object is effectively flat with-respect-to-time, it will be called dynamically flat or d-flat for short. (This is probably not the same thing as c-sharp.) The d-flat process will be illustrated next.

It can be argued that when we consider a 3-D sphere with-respect-to-time we are actually considering a 4-D hypersphere. This is true, but only when considered simultaneously in its entirety. This is the main reason the universe can be considered a "space-time continuum."

To distinguish between the 4-D universe and the dynamic 3-space, the former will continue to be called "space-time" and the latter denoted as "space/time." This could be read either "space in time" or "fixed space" at any (particular) time.

Since space-time is just another asys (axiomatic system), it is incorrect to say that this is the only complete and correct view of the universe. Since the strongest heart for the space network is not space-time (see discussion above and in "***Relativity Concept***" below), then it is not even correct to say that the universe is a space-time continuum.

However, it is correct to say that space-time is 4-D, whereas space/time is always 3-D with a fixed (current) value for the time variable t. Since all four dimensions of space-time are seldom used simultaneously, it is simpler to consider the universe to be a "dynamic 3-space." The effect of relativistic velocities on this "space/time" will be discussed in "***Relativity Concept***."

This book will generally refer to dynamic space as D1, D2, or D3 depending on the dimensions of the space.

5th Uni-Verse

"If you go on a
Hyperspace vacation;
Always guard against
Hyperventilation."

6. Cosmic Flatness Concept

Vacuum Energy

Since the late 1970s, cosmologists have known that the universe is very nearly d-flat. (Its geometry is almost Euclidean). "Galactic contraction", however, precludes visual expansion from the precise dynamic flatness value calculated for space network acceleration.

This would be troubling if events in the universe occurred by chance. A flat universe is extremely unlikely to ***occur*** by coincidence and even less likely to ***persist*** by chance. There was previously no concept to explain the cause of cosmic flatness and no concept of how it could have become d-flat except for inflation. There was previously no understanding of what caused the run-away expansion to inexplicably end precisely when the universe became d-flat!

Current (Dynamic) Flatness

Nor was there any understanding of what is causing the universe to remain d-flat. "***Cosmic Flatness Concept***" can now finally answers these questions and more.

In 1998, observations of the distribution of type Ia supernovae astonishingly suggested (due to their extreme absolute magnitude and hence great distance of visibility and ancient existence) that the expansion of the universe must be accelerating. This acceleration is sometimes called, "cosmic acceleration." **MU** will posit an interesting relationship between cosmic acceleration and the epoch of inflation along with *narratives* for both.

Stability of expansion discussed in "***Expansion Concept***" above partially explains the requirement that the geometry of the universe remain flat during expansion. "***Cosmic Flatness Concept***" will use the geometry of space to predict an exponential doubling of the linear dimensions of the universe.

If the universe is not precisely d-flat and does not continue to remain d-flat by exponential doubling then this theory is incorrect.

We will assume all of the results of "***Expansion Concept***" above. We will also assume that some natural mechanism serves to drive the universe toward time-related, dynamic flatness.

This drive indirectly caused the epoch of inflation. The logical mechanism that might be expected to cause flatness would be the motion of some type of object through the space network. The leading contenders are …

1) Massive objects (e.g. galaxies) or …

2) Photons and other monopons. Monopons will be discussed in "***Mass Concept.***"

Since no massive objects yet existed during the early creation process when inflation first occurred, then this is not a good candidate to explain inflation. Thus, it is probable that motion of electromagnetic radiation through space led to the "stability" necessary for dynamic flatness.

This is simplistically mindful of water pressure in a fire hose that stiffens and straightens the hose.

Some might worry that radiation could not be the cause of inflation because the early universe was opaque. However, opacity was caused by interaction between free electrons and electromagnetic radiation. This is not a problem since the epoch of inflation was over before (or ended soon after) electrons were formed.

There is yet a possibility that massive objects may have eventually had a further refining effect on the "global acceleration" of today. This would only be likely to occur if:

- Opacity eventually became a problem, or …
- The size of the universe eventually began to exceed the capacity for photon pressure alone to suffice to maintain stability.

There are currently no *narratives* for these eventualities. Note that the mass of the universe is not a problem since, as we will soon see, "mass" only affects "galactic motion" through the space network.

Furthermore, even though expansion is global, there is no compelling reason the cause must be global! If it is a local event, it is only necessary that it be persistent and consistent for the effect to be global. The property of global invariance, for any reason, will be named "global consistency" and its adjective will be "globally consistent." They can be shortened to "globacon" and "globaconic," respectively. Parochial c is an example of a globacon and universal expansion is globaconic.

Transflat Model (Figure 2 below)

For the purposes of this discussion, we will choose the simplest applicable model of real D3 to make the following *flatness explanation* more visual or "transparent." We will call it the "Transflat Model."

The Transflat Model is original to **MU**. It seems to represent dynamic flatness well. For this reason, the model will not be copyrighted separately so that others, who wish to, may use it in their writings … with grudging attribution, of course. However, please use your own drawing. People will begin to understand the model (and the concept) better as they see different demonstrations with it, using different versions of it, with different sizes and different labeling.

The corresponding curves from A to B and A′ to B′ in the Transflat Model (below) each represents the same entire pathway but at two different times. Co-temporal curves will be called "contours".

Figure 2

D-flat, Plane Sector,

Section of the dynamic

Transflat Model

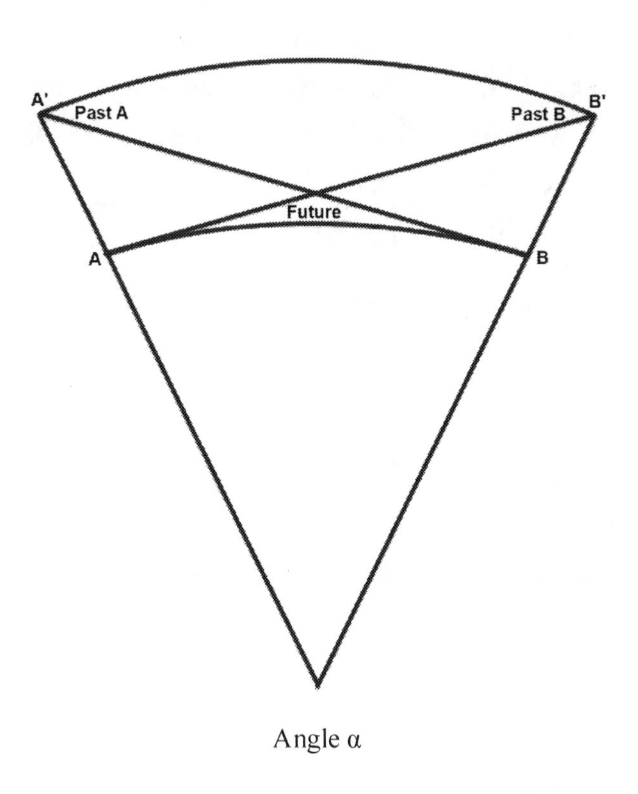

Angle α

The straight lines from A to B′ and B to A′ represent the trajectories of two photons at the exact time they passed through. Such trajectories will be called "treks." Simultaneous curves such as these two will be called "matched curves" or "matched treks". All four paths AB, AB′, BA′, and B′A′ represent the same path but at different times and under different circumstances. Such paths will be called, "matched plots".

The simplest understanding occurs when the matched plots are thought of as photon treks <u>along the surface of the universe</u>, or at least a constant, proportional distance "below" the assumed roughly-smooth surface (such as 10%). It is important to understand that this discussion is not about the "surface" of the universe. The distance below the "surface" could be any percentage, hence the principle being discovered applies to any time or place in the universe.

Caution! Do not be too quick to suppose that motion along the surface of the universe must be a curve! Great question, however. This is exactly what we need to discuss. Then you can decide for yourself how this model behaves, and ultimately if the Transflat Model accurately models universal behavior.

Time is not absolute in the Transflat Model; it is increasing exponentially upward. This is a 2-D section of D3 (a dynamic 3-space).

To maintain linear integrity, time increases radially from the bottom vertex upward as the universe expands. As one light beam travels from A to A′ and another is traveling simultaneously from B to B′, they are not traveling through time. All motion is done in real time.

Every photon in the concept is always seen at its "current" position. Two different positions for the same photon must represent different times in this imaginary "time exposure." Conversely, every point represents many different photons at different times.

This construction causes expansion to seem constant. It does complicate calculations, however.

No matter <u>where</u> you are in the universe (the actual universe, not this diagram) and no matter <u>when</u> you are there, every direction you look always reveals a vision of the past. Thus, every light beam that we can see as we look back into the past is always traveling toward us in its future.

Please understand, a generic light beam may be traveling in any direction and its source may be anywhere, but a light beam must travel in a general direction toward us in order for us to see it (when and if it arrives where we are).

The dynamic geometry in the Transflat Model (with respect to time) is assumed Euclidean for the purposes of this illustration. Thus, this should not be considered a proof, but rather an illustration of flatness. The light beams are traveling, in real time, toward a time when the universe will be larger.

As the beams travel (at light speed), the universe is expanding toward the exact size it will be but now, for this diagram, already is. This is evidenced by the fact that scaled down straight lines remain straight and thus appear straight in the diagram when seen in real time. Hence, even though the universe is instantaneously closed, its internal photon motion (with time) is always flat.

One other feature of the diagram: when hovering near A or B, A′ and B′ will be the future locations of A and B (respectively) in the direction as labeled. Thus, when hovering near A′ or B′, A and B were the same two points as A′ and B′ (respectively) as they were in the past from the direction as also labeled.

Static Flatness

There seems to be a second, more drastic, alternative called "static flatness" within "***Cosmic Flatness Concept***" that could help explain the nature of currently observed flatness in the universe by freezing it in time (at Earth's current millennium). For the purpose of this discussion, we will assume the earth is midway between the central region and the fringe region of the expanding universe.

Consider these changes:

- Inflation to flatness (Stage #1) is occurring, but only on a managed scale and only on the very young outer fringes of the universe (as seen from the earth).
- D-flatness (Stage #2) is being maintained by controlled expansion (where we are) between the outer fringe and the older central portion of the universe.
- Near-flatness (Stage #3) in the older central portion of the universe has somehow been achieved to the necessary tolerance.
- Flatness (Stage #4) in the future will be maintained indefinitely when allowances are made for time differences. This will begin near the older center of the universe and progress outward.

Are you missing something?

Consider the above alternative carefully before you read on **...** please.

You probably discovered for yourself that static flatness is somewhat simplified, but is instantaneously similar to dynamic flatness if we consider ourselves in the midrange portion of static 3-space at stage #2.

Stage #1 could be viewed from where we are by use of look-back time if we were able to see far enough back in time to the epoch of inflation. [We are in the central portion because this "now" region is the boundary region between the past and the future (stages #3 and #4).]

We now see that the vertex point, angle α, not so much "is" the center of the universe as it "was" the <u>entire</u> universe, O_0, at the beginning of time.

The 4th stage of expansion is not pictured here. Just picture it in your imagination.

Among other options, it is also possible to reverse the locations of past and future within the hypersphere. Instead of placing yourself in the past at the center of a hypersphere, place yourself on the outer fringe of our model so that all images are converging in the future as they arrive through time from the past.

If you have any difficulty in understanding how to get from the center of space to the fringe of it, just remember that dynamic 3-space is just one version of hyperspace. You need not <u>travel</u> from the center of the sphere to the edge; simply consider the dynamic 3-space to be 4-space and "hyper-rotate" the 4-space model, or "hypertate" it for short.

It is all about perspective. In this case, we will give it the adorable name "hyperspective." (Note: It does not matter that the inner spherical boundary of a 4D hypersphere seems much smaller than the exterior spherical boundary.

This is an illusion caused by forcing a model of a 4D hypersphere into 3-space. You can imagine that it looks smaller because it is further away. (You can look up the analogous "tesseract" on line.)

(Due to the symmetry of hyperspace, it is possible to mentally hypertate your position in space without changing space, itself. Alternatively, you can just change your coordinate system if that sounds easier. To each his/her own.)

Now back to our Transflat Model. If it seems that even our model expansion is accelerating from #1 to #2 to #3, just remember that the universe is larger at stage #3 than it was in look-back time at #2 and especially at #1. The expansion is accelerating exponentially; as we look further inward, we are seeing an earlier, smaller universe.

Continuous D-flatness

Imagine as the universe expands that two light beams are traveling between points A and B (on the Transflat Model). One is traveling from point A to point B. The other is traveling simultaneously from point B to point A.

Since the distances are the same and the beams are traveling at the same speed c (to the appropriate scale) then they will both reach their destinations simultaneously. By the time the light beams A and B have reached points B and A respectively, point B has expanded with the universe to its new position called point B′ and point A has expanded with the universe to its new position called point A′.

Warning! This is not like aiming a gun. It is not necessary to "lead" the motion with your aim. If "laser point A" were fired from a laser toward target B, it would not be necessary to aim at point B' because the target is not "moving"; it is actually everything in space that is expanding. Notice that the light beam mentioned above that is traveling from A to B could actually be the "laser point A" just mentioned.

In traveling from A to B′ and B to A′, curve AB has become curve B'A' while curve BA has become the equivalent curve A′B′. However, in doing so the first light beam has traveled the (straight) path of line AB′, while the second light beam has traveled the (straight) path of line BA′.

(Hint: a scaled down straight line continues to appear straight only in a flat geometry because similar triangles only exist in the flat universe of Euclidean geometry.)

1. Looking toward the future from point A, we see that the line AB′ is initially tangent to curve AB at A.

2. Looking toward the future from point B, we see that the line BA′ is initially tangent to curve BA at B.

3. Looking toward the past from point A′, we see that the line A′B was ultimately tangent to curve A′B′ at A′, which is the same vision as #1 above from a later point of view.

4. Looking toward the past from point B′, we see that the line B′A was ultimately tangent to curve B′A′ at B′, which is the same vision as #2 above from a later point of view.

Thus, both d-flat lines are true representations (with respect to time) of the (expansion dependent) linear light path between A and B.

Note that the light traveling from B to A′ is symmetrical to the light beam that travels from point A to point B′. When viewed with no expansion, both d-flat lines appear to be curve AB in the earlier universe or curve A′B′ in the larger universe. (Intermediate arcs can easily be added.) Thus, without continued expansion the universe would be closed.

It is easy to see that lines AB′ and A′B are both tangent to arc BA. (This is important since, in a closed but expanding universe, photons are expected to travel linearly.)

However, as mentioned previously, the two lines AB′ and A′B, and the two arcs AB and A′B′ are all the same line from different hyperspectives.

Therefore, it is clear that not only are AB′ and A′B both tangent to arc A′B′ but they are all the same line. From this hyperspective, it is clear that arcs AB and A′B′, which represent the curvature of the universe, are also dynamically identical to straight lines.

This reasoning, apart from the diagram, verifies geometrically what has already been shown logically. It will next be shown algebraically that the universe is d-flat.

It would make this figure considerably more complicated looking, but it would also be possible to break down the motion from A to B (or from B to A) into short segment-like arcs that would all be parallel to each other and tangent to the (current) surface near which they would be traveling at that time (not shown).

We could carry this one-step further and draw similar but shorter arcs that would be so short that they looked like points. The figure would again be much like Figure 2 except lines AB′ and A′B would then resemble dotted lines.

It should be mentioned one more time that the d-flatness discovered here is not just on the surface of the universe. Since the starting place for this study was anywhere in the universe, at any time, then it applies anywhere in the universe at any time.

Author's note: Now, if that was not fun then you just did not do it right! I know. Some of us enjoy simple pleasures, but if you did not quite get there, please try again. Success is its own reward!

Network Expansion Rate

Notice that the pathways on the Transflat Model have no indication of size. This is because the shape of the figure would be the same regardless of the actual distances. All distances would be in the same proportion in any case.

Also, notice that galactic contraction has no effect on network expansion. Motions and densities of galaxies are not factors in these calculations.

The only requirement for d-flatness is that the proportions of the space network, as illustrated by the Transflat Model, remain constant. This includes the requirement that motron stability remains constant so that light speed remains constant.

The vertex angle of the sector is called "angle a." or "<α." The measure of <α has two ramifications.

1. The corresponding size of the sector (pie wedge) is affected by the measure of <α.

2. The relative thickness of expansion area A-B-B′-A′ is affected by the measure of <α.

Please notice that changing the measure of <α would require a different diagram. The question is, “What happens to the relative size of the expansion area when the angle changes?”

It is intuitive that (in 1 above) decreasing the measure of <α will decrease the size of the sector (the area of the “pie wedge”). This is understood by anyone who has ever sliced a pie. It is not quite so intuitive that (in 2) decreasing the measure of <α will also decrease the relative expansion width (the thickness of the “pie crust” at the top between arcs AB and A′B′).

However, the following discussion deals with relative decreases, therefore the measure of <α will be irrelevant. Its effects will be cancelled out by the use of indices. To see this, note that the measure of <α is arbitrary in every calculation.

Pie Crust, Indices and Expansion Factors

(If you are uninterested in the reason (ii) is true, skip this section.) Otherwise, notice that (all else remaining the same) a decrease in the measure of <α causes a decrease in time since light is traveling from A to B at constant speed. Also, remember that the expansion of the universe is not constant, but accelerating.

Remember that time in the Transflat Model is increasing exponentially upward. To maintain the same relative area of “pie crust” to “pie” would require a decrease in the expansion time upward in the Transflat Model from exponential to constant, i.e. a smaller piece of a smaller pie. You do the math. For a hint, please see the next paragraph.

If we looked at the same relationships (the Transflat Model with the same measure of <α) after all dimensions had been multiplied by a factor of f, the figure would look the same. This seems to imply that, since all distances have been multiplied by factor f, then the network expansion rate must be uniformly exponential for the particles to arrive on time.

As an aid in remembering the significance of this acceleration, space network acceleration will henceforth be referred to (with capital letters) as "SNA." We will take a few minutes to prove that this is so. Let size $S_1 = 1$.

The path from A to B′ is being traveled by a beam of light. If a beam of light travels from A to B′ with light speed c for a length of time $T_1 = t$, then the length of AB′ = c · t. (In general, distance d = speed · time and c = constant light speed.) While the beam travels from A to B′ the universe is expanding from B to B′ also in time $T_1 = t$.

Now let new size $S_2 = f > 1$. If the size of the Transflat Model is multiplied by factor f, then the new AB′ could be called A_1B_1' for distinction. Since $A_1B_1' = f{\cdot}AB' = f(c{\cdot}t)$ then $A_1B_1' = f{\cdot}c{\cdot}t$. Thus, the universe has expanded from B_1 to B_1' in time $T_2 = f{\cdot}t$.

Since light speed c is constant, then the time factor, t has been multiplied by f to get, f·t. Therefore, since the figure proportions have not changed, all expansion now takes f times as long when the distances are f times as far.

If it takes f times as long to expand f times as much, then the expansion rate has not changed. We will call this expansion index $r = T_2 : T_1 = f{\cdot}t : t = f : 1$; similarly, the size index $d = S_2 : S_1 = f : 1$.

Notice that f is actually a dilution factor. When the expansion is divided by the size, it has the effect of distributing the (linear) expansion over the entire (linear) size.

We can now determine the expansion index $\Re$ of size index δ compared to expansion index ρ when $\rho \neq 0$.

$\Re = f : {}^{1}/_{f:1} = 1$ where d and f are both positive.

Alternatively, $\Re$ could be calculated $\Re = (S_2 : T_2) / (S_1 : T_1) = f : {}^{ft}/_{1:t} = 1 : {}^{t}/_{1:t} = 1$.

Thus, the ratio of linear expansion factor, f to distance expansion factor, f is f/f = 1.

Furthermore, the area expansion factor is the square of the distance expansion factor, or f^2 and similarly the volume expansion factor is the cube of the distance expansion factor, or f^3. If this is obvious to you, then please skip the next paragraph.

Given that the new length $l_2 = f \cdot l_1$, the reason the area expansion factor is the square of the distance expansion factor f is that area is equal to length l times width w. If the new length $l_2 = f \cdot l$ and the new width $w_2 = f \cdot w$, then the new area $A_2 = (f \cdot l) \cdot (f \cdot w) = f^2 \cdot (l \cdot w)$. We now see, the new area $A_2 = f^2 \cdot l \cdot w = f^2 \cdot A_1$, where A_1 is the original area $l \cdot w$. Similarly, the new volume $V_2 = f^3 \cdot (l \cdot w \cdot h) = f^3 \cdot V_1$, where h is the original height and V_1 is the original volume $l \cdot w \cdot h$.

Space Network Acceleration vs. Galactic Contraction

"***Visible expansion***" is the well-known motion of galaxies within the space network. This motion of galaxies and cosmic protuberances is the large-scale, measurable motion of the visible universe. It is not exponential.

"***Galactic contraction***" is the component of network expansion that is totally controlled by gravitation. It is negative expansion within the space network. As such, galactic contraction is not responsible for the expansion of the universe but actually works against it. Galactic contraction is the difference between the currently undetectable network acceleration and visible expansion.

The concept of network acceleration as a motion is much different from galactic contraction (beyond the ± direction). "***Space Network Acceleration***" (SNA) is an original hypothesis being first introduced in the previous "***Expansion Concept***" and now being used in the current "***Cosmic Flatness Concept.***"

These concepts are useful in the investigation of relative differences between the mechanics (different causes) of the two components of visible expansion. This will be done by answering 3 critical questions.

First Question

Q1: During inflation, exactly what was it that was expanding?

A1: There were 2 phases of universal expansion. Phase 2, called cosmic acceleration, will be discussed shortly.

Phase One Expansion (Cosmic Inflation)

Phase 1 is called cosmic inflation. There was initially nothing to expand in phase 1 except for the singularity of dark energy (fabrons) of the space network itself.

After the Big Bang, but still during the first second, inhomogeneities within the space network coalesced into globules of elementary particles. These joined (continued with) the expansion even as the globules were forming within it.

During this first second, microgravity began to form but this was well before galaxies came into existence. At this time, there was no (or very little) galactic contraction.

The mergers of inhomogeneities into globules might be effectively similar to micro-gravitational contraction. However, all positive expansion (inflation) was via the "network."

There are 2 versions of what may have caused phase 1 inflation:

1. For some reason, "the universe entered an unstable, high-energy state called a 'false vacuum'."
2. Dark energy expanded during inflation to create the space network (of energy). The photon need for stability (photon pressure), flatness, or both may have caused inflation. Singularity reduction may have been the vehicle that carried it out. There will be more on this later.

Dark energy of the space network is obviously not bound by the rules (e.g. speed limit c) that have restricted motion within the space network since it formed. Any incidental space curvature (beyond d-flat maximum curvature) can almost instantly be remedied by appropriate expansion of the space network. Parochial c will be discussed in "***Relativity Concept***."

The initial universe was obviously closed. A (flat) Euclidean 3-space is considered infinite. Thus, it would have been impossible, in a finite amount of time, to make the finite universe flat. No matter how long or how fast you uniformly expand a sphere, it is still a finite sphere. In almost no time, however, the universe probably became "dynamically" flat.

The four properties: flat, infinite, homogenous, and isotropic will be called the "dynamic properties." This is because they require (continued) expansion for their initiation (and for their survival).

[Space-time was proposed (in part) as an early mathematical alternative to the then yet undiscovered dynamic flatness being initiated here.]

Phase 2 Expansion (Cosmic Acceleration)

The phase 2 versions of all four dynamic properties continue to this day. Phase 2 expansion is called "cosmic acceleration"; expansion has now slowed to a proportional (exponential) rate.

The space network expanded rapidly during the first minutes after the Big Bang. Almost undetected at that same time, tiny warp fields were growing from small "inhomogeneities" to somewhat larger "protuberances" by acquisitions and mergers between them.

This type of increase in size will be referred to as "growth," not as "expansion." These massive objects eventually grew into star clusters and galaxies.

At least some of the warp fields (probably most and maybe all) were free floating within the space network. This growth is part of the clumping process mentioned earlier.

Clumping increases the total mass of individual galaxies by gravitationally bringing more mass into the galaxies from the surrounding space while gravitationally decreasing their volume. (This is allegorically like squeezing out excess "space".) Clumping automatically increases the volume of the "empty" space between galaxies by (figuratively speaking) transferring to it the empty space that was squeezed out of the galaxies. This expansion of the universe is apart from (in addition to) the continued expansion of the space network itself.

Due to the presence of galaxies and other matter within the space network, there are two categories of cosmic expansion possible.

Category B is the name we will give to expansion that involves increasing average distances (and volume) between galaxies.

Similarly, Category W expansion would involve increasing average distances between objects within galaxies along with increasing the average size (volume) of the galaxies themselves. There are four possible combinations of these two types of expansion.

Combination 1: Both categories B and W are involved.

Combination 2: Only category W expansion is involved.

Combination 3: Only category B expansion is involved.

Combination 4: Neither category is involved.

As massive objects grew, there was no reason and probably no way (see discussion above) for the known cosmic expansion to have occurred without category B expansion. This eliminates the possibility of either combination 2 or combination 4.

Due to the clumping problem, however, the universe of protuberances probably expanded with the network but not as the network.

"***Expansion Concept***" will therefore assume that cosmic expansion involves category B (between) but not category W (within) expansion.

This corresponds to only combination 3 above. This property of galaxies will be called "external expansion." This condition must also have prevailed during their formation from independent protuberances.

Growth of galaxies is limited to their mergers and acquisitions. This is not necessarily a dynamic situation since stars and even galaxies are also continually being destroyed and scattered to rejoin the raw material destined for new mergers and acquisitions.

This allows for very little significant change in general galactic density so that galaxies, as they mature, can continue being considered independent protuberances. Because of this and their fixed general density, warpon intensities (within galaxies) must also be fixed.

To put it another way, the primary distribution within the universe that is not a fixed constant as globules coalesce into galaxies and the universe expands, is the generally widening distribution (spread) between galaxies within the universe. However, there is compensation here as well with continual introduction of new galaxies of all sizes and shapes. It is not yet known if studies on this final component are conclusive. The question remaining is whether new galaxies are being formed at a sufficient rate to compensate for destroyed galaxies as well as to "populate" the new space created by universal expansion. For the remainder of this discussion it will be assumed that the universal density of matter is diminishing. This will be called, "universal wasting."

However, photon pressure along with D-pushing are likely responsible for *motron* stability D. The notion of constant motron stability is called the "Motron Stability Conjecture." For more detail on this conjecture, please see "***Advanced Speculation***."

Galaxies and other globules are free to move around under the influence of both their momentum and their gravitation as the universe continues to expand to this day. Thus, SNA and stability apply to all space of the space network.

However, the warp space within the space network features fixed density and fixed size (except for acquisitions) of the current contents. This all implies that cosmic acceleration is only occurring between galaxies, whereas constant motron density is likely continuing everywhere in the entire space network in spite of global acceleration. For more detail on the seemingly impossible claims in this paragraph, please see D-pushing in "***Advanced Speculation***."

Second Question

Q2: Since a very large universe is not necessary for dynamic flatness and a smaller universe need not expand as rapidly as a larger universe, what caused the universe to become so large before inflation ended?

A2: Besides the required exponential expansion rate, there are two subtle requirements for d-flatness. Before d-flatness is possible, parochial constancy is necessary. This requires the universe to be stable over time with uniform motron density D. The "matterfacation" process (energy converted to matter) must first convert all energy being mattified from singularities to singletons.

Therefore, expansion is necessary to accommodate total-singularity-reduction. Fixed density also requires "D-pushing" stability. D-pushing and other matterfacation requirements will be discussed in "***Advanced Speculation***" below.

Third Question

Q3: What type of expansion remained after universal inflation ended?

A3: It is true that the epoch of inflation ended, but there is no evidence that inflation-like-expansion has ever ceased. As we have seen, cosmic inflation was needed to make the space network flat. This was probably accomplished as mentioned by a total singularity reduction of all energy being mattified. Having achieved dynamic flatness in this way, the urgency of expansion was greatly diminished.

Continued slow-motion inflation in the form of SNA must continue today to keep the space network d-flat because photon pressure is still present. This continuing inflation will be called "margin inflation." One of the reasons that continued margin inflation is considerably tamer than the original epoch of inflation is that total singularity reduction was likely a one-time necessity.

Extreme inflation is no longer necessary since all warpons are probably singletons and fabron density has reached the required value of D for motron stability. One side benefit of margin inflation is the ability it gives D-pushing to maintain motron stability.

Increased size does not change the type or the shape of universal curvature. The universe is probably an expanding sphere, oblate sphere, or spheroid. The increase in size however does tend to decrease the "density" of curvature by distributing it over greater area and greater volume. (It is likely that gravitation and photon pressure have formed the universe into a sphere similar to the way surface tension and internal pressure form soap bubbles into spheres.)

Mathematically speaking, the sphere embodies the maximum efficiency of surface area vs. volume. Furthermore, there are far fewer complications to consider if we (at least temporarily) imagine the universe to be spherical.

It was shown in "***Expansion Concept***" that the required rate of expansion that is needed for d-flatness is proportional to its size. As the universe expands, galactic contraction may become less important due to loss of material density that decreases the effect of gravitation. Galactic contraction therefore probably has decreasing effect on the rate of visible expansion, which is the only motion that is directly measurable.

Visible Expansion

The net result of both components of expansion with respect to size is primarily caused by exponential SNA. To be more specific, the rate of SNA that doubles the linear scale of the universe is constant with respect to size. This means that the network rate of linear expansion is a little greater than the visible expansion rate. The difference is galactic contraction.

Galactic expansion is negative, so it can be called "galactic contraction," especially near the unbalanced gravitational fringe region. Although the high expansion rate of the network is now greatly reduced from that of its micro-epic epoch, the universal loss of material density has been prodigious. This loss has deprived galactic contraction of sufficient gravitation to allow it to be as much of a counter-balance to SNA.

Galactic contraction has no effect on flatness because SNA is completely responsible for flatness and because galactic contraction only affects the visible expansion. Galactic contraction is nonetheless important since it does affect galactic distribution within the universe and visible expansion is the only expansion that is directly measurable currently.

It was the network, having expanded during inflation, which caused the universe to become d-flat. Ever since that time, SNA has been fighting the finiteness of space to maintain the d-flat space network.

Notice that the requirements for d-flatness do not consider "warp field density." It is totally based on geometry. Density is but a partial measure (or symptom) and not a driver of flatness.

As the universe expands point B to location B', <u>what if irregularities prevents point B from arriving at location B' in time</u>? This would mean that the universe was not quite d-flat. Therefore, the calculations for d-flatness are actually calculations for the critical expansion requirements for d-flatness. This, therefore does not constitute proof that the universe is d-flat.

However, if it is not d-flat, then the amount by which the expansion rate falls short will be called the "expansion deficit." If the deficit is not zero then the deficit will likely be increasing exponentially. It seems extremely unlikely that the geometry of the universe could continue to be close to Euclidean geometry under those conditions.

"***Cosmic Flatness Concept***" will therefore assume that the universe is effectively d-flat. If the deficit is not zero, then it will be assumed decreasing, constant, or increasing at some sub-exponential rate. There is yet no *narrative* for this possibility.

Since SNA has been unrecognized until now, the current approximation of visible expansion is our best measure of cosmic expansion. The visible expansion rate is equal to the SNA rate minus (plus the negative) galactic contraction rate. Measurements, calculations, or estimates of any two of these rates will eventually produce information about all three rates. (This requires that those doing the determinations know and care about all three.)

Even if network expansion is confirmed to be exponential, the study of the effect of critical density on galactic contraction is still important to allow for reconciliation of network calculations with direct observations of visible expansion.

Conclusion

"***Cosmic Flatness Concept***" explains cosmic flatness in the guise of d-flatness. D-flatness then explains the process and the essentials of visible expansion called "global acceleration."

Galactic contraction is diminishing, but it is only important in this regard because it can be estimated and because of its effect on visible expansion. Visible expansion is important because global acceleration has been tentatively verified and may someday be calculated to a discriminating tolerance.

Measurements and calculations show that visible expansion is not only accelerating, but is accelerating exponentially while compensating for gravitational contraction. This tends to confirm "***Cosmic Flatness Concept***." *Narrative #13* then removes it from the realm of regality.

Imports of the "Cosmic Flatness Concept"

To summarize this discovery, the theoretical SNA expansion rate required for d-flatness is proportional to the size of the physical universe. This would make it exponential.

The predicted acceleration of net expansion seems to be confirmed by new cosmological findings that the universe appears to be accelerating. The fixed network-expansion-correlation-index of 1 has not yet been independently confirmed.

Since the space network is not yet detectable, then it is impossible to separate out its expansion rate. Visible expansion superimposes galactic contraction on top of the invisible SNA

"***Cosmic Flatness Concept***" has now shown that the linear expansion rate of the network is directly proportional to network linear size. Therefore, the SNA will theoretically always* be increasing at an exponential growth rate.

According to some of the latest studies, when SNA is added to gravitational <u>deceleration</u>, the <u>net</u> result is <u>visible expansion (acceleration)</u>. These two components are not visible. Cosmologists can only measure the visible expansion. This simply means that the gain in acceleration from the network also compensates for the loss of expansion in the galactic contraction. This allows SNA to maintain a fixed exponential rate.

** The word "always" above requires qualification. At some point in the future the maximum expansion of the universe at which the value of D can be maintained will probably be reached. Besides this limit, there could also be a limit on how fast the universe can expand.*

If humans survive (for hopefully a very long time) to see these expansion limits, many possible fates await us. Most of them are unpleasant. This *narrative* must (happily) be left for future development.

Universal C

There have been many discoveries in recent centuries that have been in conflict. Not the least of these ponderables has been the question as to how the speed of light could be constant within a universe that was expanding at a rate that did not seem constant. Since the density of the "fabric" of space was assumed to be variable, then the constancy of light speed was thought to be solely a property of photons.

Now that we know about the stability of space and the dynamic flatness of its geometry, the speed of light is a whole new concept. To pursue this concept, we must learn more about how warp fields might effect motion.

Early Warning

Due to the relation between c and SNA, the first sign that the universe might be approaching the expansion limit would likely be the appearance of slight fluctuations in the measured value of c. It seems likely that, even though the problem has no possible solution, someday an early warning system may alert mankind to the impending demise of stability.

6th Uni-Verse

"Local phenomena can apply globally
If they are persistent and consistent."

7. Ruffle Concept

Warp Status

All warp fields are very much alike except for size, shape, and the type of motion being exhibited with respect to the space network. However, an understanding of warp requires recognition of the relationship between particular objects and their individual warp fields.

The simplest classification of warp is "internal warp." This classification refers to an objects own warp. Its status, whether moving or stationary is the same as that of its massive object (an object possessing measurable mass) that is always with it.

It should be pointed out that the adjective "internal" does not imply that the internal warp is contained within the object. The internal warp of an object refers to the (in-house) effect of the warp upon its own object.

Symmetry

When an object is at rest, the energy of its 3-Dimensional internal warp field is balanced symmetrically around the center of gravity of the object. This "central symmetry" of the object's mass to its center of gravity is a clue to the relation between internal warp and mass.

Keep in mind that the object, itself, can have any shape. The description "spherical" applies to the balanced warp of a particular element (point) of the object. The symmetrical (balanced) internal warp of a particular element of the object will be called its "underwarp." It is a repository of potential energy.

The overall shape of any larger warp field would likely map the mass "footprint" of the object. This will be a handy tool if astrophysicists someday learn to detect and measure or even "image" warp fields in space. This seems likely, due to the mass involved.

Regardless of whether a mass footprint is calculated or "imaged," the footprint will be called its "mass image." The maximum (average) density of the mass image will be located at its center of gravity.

Numerous minimal and maximal intensities are possible. These could be called "gaps" and "spikes" respectively. When two or more objects approach each other for a possible encounter, their respective mass images could eventually merge to produce a "mass collective."

This collective will increase in density as the objects continue to approach. If the objects merge, the mass collective will become a new "single mass image." The transformation from a mass collective to a mass image could be fascinating to witness. "Collisions" and near misses might be at least as interesting.

What Are Ruffles?

1. Concentrations of unbalanced warp will be called, "ruffles."
2. Ruffles of linear internal warp are called "afterwarp."
3. Rotational internal warp produces "anglewarp."
4. Revolutionary internal warp produces "orbitwarp."

<u>What is a Monopon?</u>

1. A photon and other similar particles probably have exactly one internal (indivisible) warpon (called a monopon).

2. It is always ruffled to the same extent.

3. Since it can never be balanced or symmetrical, it will always produce motion.

RUFFLES

Figure 3

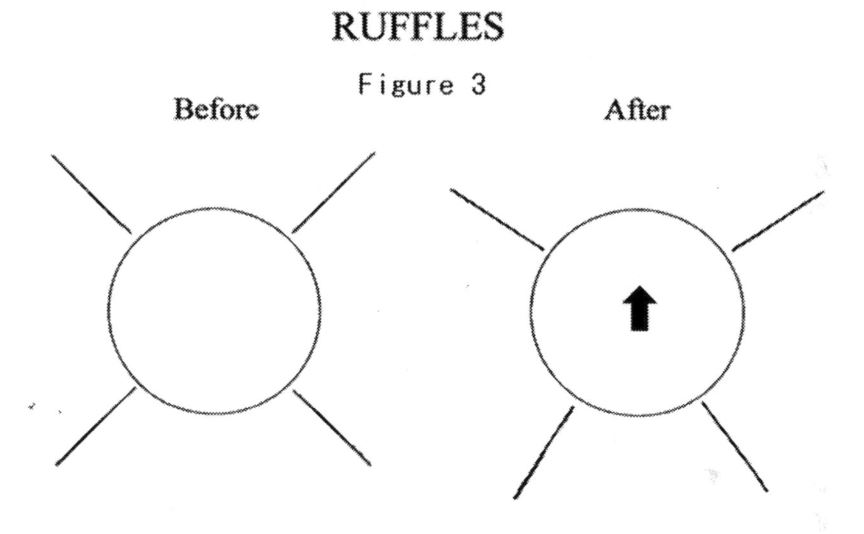

What is Parochial Constancy?

1. The parochial (local reference frame) speed of a photon must always be the same (constant).

2. Parochial constancy of c will be called "parochial c" for short.

This *narrative #14* partially explains constant c, beginning the removal of parochial c from the <u>realm of regality</u>. Since there has never been an acceptable *narrative* for constant c that includes alien c, then the "constancy of alien c" must remain in the realm of regality.

It is easy to understand why ***universal*** c may have been considered constant when there was no *narrative* for constant c and no understanding of the difference or the need for the distinction between parochial c and alien c or how this distinction might affect the variability of c. Parochial c will be discussed further in "***Relativity Concept***."

There are three *fundamental* types of ruffling being introduced in "***Warp Concept***:" (Linear) revving causes afterwarp and gravitation. Angular revving causes anglewarp and precession. Orbitar revving causes orbitwarp and nielucion. These will all be discussed soon.

Revving (Warp Unification)

Due to inertia, any change in motion produces a change in the ruffle pattern, which requires acceleration. Warp displacement (motion) can be caused by electromagnetic and mechanical forces. These will be called general forces. "Manual motion" is caused by throwing, hitting, pushing, pulling, or etc.

Any motion that ends with either (a) projection or (b) throwing and releasing while still in motion [like (i) firing a gun or (ii) throwing a ball] will be called "revving." Twirling or flipping a disk will be called "spin-revving."

When an object is revved (or spin-revved), the object's warpons are displaced or stripped backward away from the direction of motion. The backward stripping of warpons is mindful of ruffled feathers. Accordingly, they will be called "ruffles."

Of course, spinning dresses and skirts, even blouses and shirts (like on a dance floor) can have both kinds of ruffles, but who knew that "flying" bullets and cannon balls (like on a battlefield or battle frigate) could also have ruffles?

Warp displacement caused by mechanical or manual motion, such as revving, will be called physical warp or internal warp. Linear revving results in the warp field being concentrated on the aft side. This aft-concentration of warp is now being called "afterwarp." Spin-revving results in anglewarp being concentrated along the (faster moving) outer edge of the rotating object. Orbitar revving results in orbitwarp along the orbital path of the satellite.

Gravitational warp will be called external warp. External warp refers to any warp enhancement that is attributable to an external revving. This will be discussed in "***Gravitation Concept***."

The Geometry of an Afterwarp

An afterwarp in space would generally be 3-Dimensional. For simplicity here, we will only consider linear motion of a "mass point" (a single point of a massive object in space) at this time. The intersection of a mass-point afterwarp with any plane of its motion containing the center of gravity will be called a "plane afterwarp" or simply an "afterwarp" as opposed to an ellipsoidal "space afterwarp."

The actual shape of the afterwarp concentration of a one-piece object is of only minor concern. However, its behavior seems to imply that a plane (mass-point) afterwarp can be represented by graduated, elliptical contours of warp density or afterwarp concentration.

These will be called "isowarps" (like isobars of a weather chart) with all leading foci at the center of gravity and with a common major axis parallel to the direction of travel.

The evidence also seems to imply a relationship between the warp rate and the eccentricity of the isowarps. If this is true, then the eccentricity is likely to be 0 precisely when the warp rate is 0 and 1 when the warp rate is 1.

With more evidence, it may turn out that an isowarp's eccentricity is simply a measure of the warp rate, or at least an approximation of it. More data collection will be needed to verify this possible relationship.

"***Warp Concept***" will assume with little consequence that each afterwarp is elliptical with eccentricity equal to its warp rate. It seems likely that anglewarp can also be represented by concentric isowarps except that they are always likely to be circles.

The turn center and direction of rotation should also be indicated in this case. The distance between concentric circles such as the "speed density circle" and the "unit density circle" might represent the angular speed.

Since gravitation, precession, and nielucion are caused by afterwarp, anglewarp, and orbitwarp respectively, then these can be represented analogously except since these all generally involve two or more objects, there will generally be two (or more) interacting figures each with their own "primewarp" respectively.

It seems likely that this will all become standardized rather quickly if some form of it were someday put into practice. (With appropriate software, it should be possible to display dimensional isowarp surfaces and even space isowarps.) Awesome! Just imagine the use of color and animation for advanced concepts … or for entertainment!

Verification of Afterwarp

There is currently no physical proof to support the concept of an afterwarp. Furthermore, there can be no proof of afterwarp until the discovery of sufficient physical evidence of a space network.

We are not without at least circumstantial evidence, however. Presentation of some of this evidence is an intention of **MU**. This is similar to the acceptance of gravitation based on circumstantial evidence alone, except that in the case of afterwarp there is far more evidence in far greater scope and variety than initially existed for gravitation.

Much of the evidence of warp fields is already familiar and many of the formulas have been previously established. The missing component has been the interpretation of this evidence.

Science seems on the cusp of the needed proofs. This is mindful of the search for gravitation waves. It might be useful for us to know exactly what it is that we are trying to prove. This would eventually require us to understand how gravity works.

That is the primary goal of "***Ruffle Concept***." The secondary goal is the development of definitions and descriptions that will help facilitate the communication of related thoughts.

Imports of "*Ruffle Concept*"

"***Ruffle Concept***" uses the concept of revving to posit the ruffling of internal warp. This results in warp displacement. Afterwarp is used to help us understand and describe inertia and momentum.

The dark force of external warp has no known direct effect on the associated symmetrical internal warp of a stationary massive object. Both warp sources seem to affect the object independently. Any nearby external warp will have no visible effect on a sufficiently constrained and sturdy object.

However, object A approaching object B traveling in the same direction will normally experience greater contra-warp from object B due to B's afterwarp. This will likely explain B's gravity assist to A.

Nonetheless, any nearby external warp will accelerate an approaching object by focusing and augmenting the approaching object's internal warp, effectively "creating" an unbalanced force called gravitation. As we know, the nearby object will experience the same effect for the same reason in the opposite direction.

The overwarp of a near-by massive object concentrates the underwarp on the far side of an approaching object. This focuses the afterwarp of the approaching object causing the approaching object to accelerate toward the nearby object. This far-side focus of warp will be called, "pressurewarp."

Simultaneously, the approaching object is doing the same thing to the nearby object. These two pressurewarps cause unconstrained objects to accelerate toward each other. If only one of the objects is constrained, this will not diminish the other objects acceleration. Each object's pressurewarp stimulates the other object's far side like-directed afterwarp.

As we have seen, afterwarp is associated with speed. Even an artificial increase in an object's afterwarp causes that object to accelerate away from the increased afterwarp. This is called a "simulated afterwarp."

A simulated afterwarp thus induces real speed. The induced velocities of both objects generate increases in the "regular" afterwarps of both objects and this produces acceleration in both objects. Acceleration is caused by force. This induced force is called the "force of gravitation."

Once we understand gravitation, it seems so obvious. However, it was not figured out long ago because we did not understand about the afterwarp. Now we can see that a warp field uses its afterwarp to maintain the speed of momentum in the host object but the pressurewarp of gravitation from the same warp field also causes acceleration in unrestrained nearby objects.

The two pressurewarps generate gravitational forces that are equal and opposite. However, the reactions of the two objects to their corresponding gravitational accelerations depends on their own masses by the equation $a_1 = f / m_1$ and $-a_2 = -f / m_2$.

In effect, the objects are being pushed toward each other with the same quantity of force. The less massive object is acquiring proportionately more acceleration, however.

When dropping a small rock, the rock is seen to accelerate toward the earth. However, the massive earth does not even seem to notice the presence of the small rock. The effect in that latter case is minimal.

"***Ruffle Concept***" has now partially explained how gravitation works. This sets the stage for the details coming up soon in "***Gravitation Concept***." However, we must first discuss the definition of mass.

8. Mass Concept

Assumptions

Every quantitative measure of a warp field is nonnegative. Thus, the "mass" of an object is simply a summation of the measures of all disjoint warp sub-fields of that object.

Components of Warp Motion

The two types of substance in space that are related to warp motion are:

(1) There is only one space network in the universe. It is flexible and filled with its energy and dark energy (fabrons), and it is relatively …

- Steady (stationary with uniformly distributed density over space),
- *Stable* (evenly distributed over time), and
- *Continuous (*permeates all of space, including all matter).

(2) There are numerous warp fields containing massive objects with their matter and dark matter (warpons).

Imagining Components of Motion

The space network can be imagined as an (invisible) "wheat" field with (visible) amber waves of grain. The visible heads of wheat represent submicroscopic particles (quanta) of dark energy that collectively swarm around you in 3 dimensions instead of the (thick) 2 dimensions of grain.

Now all you have to do is make them smaller so you can see more of them and then steadily smaller until well beyond the point where you can see none of them (even microscopically) but you know they are still there.

Warp fields, however, are "lumpy" with each "lump" being matter or an elementary particle associated with the particular warp field bound to it. It is unclear whether we should consider that warp fields bind the matter together or vice versa or both.

To visualize warp fields, convert the space network of particles into a representation of clusters of sub-microscopic points of dark energy each containing one submicroscopic particle of dark matter.

Are you making it (and yourself) small enough? In this wonderland of contradictions, where are Lewis Carroll's mushrooms when you need them? (AKA Charles Dodgson, British Author of **ALICE'S ADVENTURES IN WONDERLAND**)

It is likely, however, that the structure of each warp field is a direct result of its (parental) relation to its associated massive object. The parental relationship will be discussed in "Matterfacation" within "***Advanced Speculation***" later.

Understanding Mass

We could spend much time discussing what we do <u>not</u> know to be true and what we do know to be <u>not</u> true. Until now, however we have had no understanding of the relationship between mass m and the corresponding force f of gravitation.

Of course, together m and f are related to acceleration by Newton's second law: $f = m \cdot g$, where g is the acceleration due to gravitation. However, until now, some of the biggest science mysteries are concerned with gravitation and mass.

What are they? How are they related to matter? How does their relationship work? What causes $f = m \cdot g$?

One of the problems is that photons are traveling at hypothetically maximum speed and thought to have no mass except for equivalent mass. Furthermore, the more massive an object is, the greater the force required to accelerate it. However, besides that, things seem to get even more massive as they accelerate.

To summarize, a specific force applied to a specific object can cause a specific acceleration of that object that increases its speed, but it also increases its mass and this inhibits its acceleration that was caused by the specific force.

Well, that explains everything!

It is assumed by many that the mass of an object impedes speed similar to the way a swimmer might be impeded by a pool full of molasses. This molasses analogy is probably a torturous interpretation of the facts.

If "***Mass Concept***" is to be believed, then all motion is affected and maintained by distortions (internal or external) in the (tacit) warp fields, which permeate all massive objects.

Before any warp displacement occurs, the warp field, called "internal warp," which surrounds every massive object symmetrically within 3-space, is balanced around the center of gravity of the object. This can be considered a merging toward the center. Alternatively, it diverges away from the center depending on how you want to look at it.

Now, that makes a lot more sense!

Symmetrical internal warp, called "underwarp" can be visualized (although invisible) as very long, submicroscopic, straight rays (resembling <u>*spokes*</u> of energy) collectively surrounding an entire massive object … every object … in 3-space. This departs completely from our usual visualization of energy as being an "invisible amorphous insubstantial dynamism."

These two visualizations of an energy field may simply reflect a difference in scale or in viewpoint. The good news is that it does not matter how we visualize energy, its behavior is what really matters. It is still a good thing if we can come up with a visualization that helps us understand relationships between mass and energy.

Constant afterwarp maintains speed. Waxing afterwarp and unhindered external warp both cause acceleration in the direction away from the greatest warp concentration.

The cells of Table 3 might help us unify the various warp fields mentioned in the table.

Warp Field Unification Table 3	**Internal Warp (speed)**	**External Warp (acceleration)**
Symmetrical (stationary)	Underwarp (dark energy)	Overwarp (dark force)
Ruffled (moving)	Afterwarp (revved impulse)	Pressurewarp (warp force)

1. Symmetrical internal warp, called “underwarp,” is associated with the “dark energy” of the space network.

2. Symmetrical external warp, called “overwarp,” is associated with the “dark force” of a nearby gravitational source.

3. Ruffled internal warp is called “afterwarp” because the revved impulse of the warp field of every linearly moving object is concentrated toward the aft side of the object.

4. Ruffled external warp, called “pressurewarp,” is associated with acceleration (caused by an external pressure) within a gravitational field.

Q: How does a warp field “push?”

A: It will be postulated that when any object, together with its warp field, travels through space, its motion (momentum) is maintained by its afterwarp in a manner truly similar to the maintenance of photon motion. More on photons later.

Q: How can a warp field behave like energy?

A: A warp field ***is*** an energy field. Energy is the ability to do work. All moving objects have the ability to do work; this kind of energy is called kinetic energy. Stationary objects sometimes have the ability to do work because of their position or status. This is called potential energy.

Q: How are mass and energy related?

A: The answer to this question requires careful wording. The simplistic answer seems to be that they are two “features” of the same essence. We have long known that there was an equivalence between mass (m) and energy (E) as expressed by the well-known formula involving the speed of light, c: $E = mc^2$.

There have been many attempts to explain or even "talk around" this relation. Some consider this the "conversion formula between mass and energy." Without "***Mass Concept***," this task is difficult. Let us try it now.

Since by definition of energy, $E = mv^2$, then an object with mass m moving with speed v would have energy $E = mv^2$. The momentum p of a photon traveling at speed ($v = c$) should be (by substitution) $p = mv = mc$ because the speed v in that case would be c.

However, it is <u>impossible</u> to convert between mass and energy except in viewpoint or semantics. We know that mass serves two distinct functions: inertia and momentum. (Inertia and momentum is physically the same thing, but we choose to define and view it through different lenses.) By our interpretation, then, every object in any system always has at least one of these. If the object is stationary (within that system), it has inertia. If the object is moving, it has momentum. However, in addition to the momentum of its motion, a moving object also retains its inertia. In both views, inertia is "resistance to acceleration," whereas, momentum is "maintenance of speed."

Momentum is caused by impulse, which can be in the form of (force · time) read: "force times time." Impulse can also provide the ability to do work in the form of (force · distance).

"***Mass Concept***" will accept that mass and energy may be thought of as the same "stuff" with two different roles. We will call it "mass/energy."

When called "mass," the mass/energy of some massive object can be thought of as fulfilling its ***inertia*** role (inertia = "resistance to acceleration") by opposing acceleration of the massive object (mass = force/acceleration). To understand this formula, note that acceleration is in the denominator. This implies that greater acceleration requires either more force or less mass, or both.

Contrarily, *molasses* opposes speed, not just acceleration (except by association). It may be misleading, therefore, to compare mass and molasses in any meaningful way. Also, notice that the mass of an object travels with the object whereas molasses would epitomize an external impediment to motion.

When called "energy," mass/energy is thought of as giving ***momentum*** to some massive object and providing it with speed. (Momentum · speed) = (force · distance) = work, whereas the ability to do work is called energy.)

(Contrarily, molasses certainly does not provide momentum except possibly in a negative way as a drain of momentum. Molasses probably does provide energy when ingested as food, but this is clearly unrelated to the concept being discussed.)

Q: What is the physical difference between mass and energy?

A: The first thing to remember is that you cannot have one without the other. When a massive object is stationary, its warp field is symmetrically balanced. Since there is no motion, the prominent feature is its inertial mass; it only has potential energy.

A tree growing out of the ground is a massive object. It receives energy from the sun, the wind, the rain, and nutrients from the air and the soil. The tree's energy from the above sources is stored as proteins and potential energy.

An unbalanced warp field of any (moving) particle, however, maintains the momentum that supports the motion of that particle. Such particles are thought of in terms of their energy. Photons of radiation, for example, are thought of as energy.

How Is Mass Related to Warp?

To understand the opposite faces of mass, consider the afterwarp. Force is needed to increase the momentum of an object. The quantity of the required force is proportional to the amount of mass, which determines the effort necessary to accelerate the object along with its warp field. This effort seems to be proportional to any measure of the warp field.

Moving the object creates afterwarp, which is a distortion of the warp field. Thus, there are at least two reasons that we can say, "Mass is a measure of the warp field distortion (ruffle)."

a. Mass in any units is a measure of the acceleration of an object (by increasing its afterwarp).

b. Mass is a measure of the resistance to slowing the object (returning symmetry to its afterwarp).

This *narrative #15* removes mass from the realm of regality.

What is dark substance?

Particles of dark substance are being called "motrons" because of their relationship to motion. Motrons that provide the fabric of space will accordingly be called "fabrons." Fabrons are considered to be dark ***energy***. Motrons that provide warp fields of massive objects will be called "warpons." Warpons are considered to be dark ***matter***.

Q: What would fabrons look like if they could be magnified large enough to be seen?

A: We seldom talk about energy in terms of what it looks like. Photons, like fabrons, also are particles of energy that are too small to be seen. Because of their brightness, photons probably contain energy. Because of their motion, photons probably contain mass.

Fabrons would probably look much like photons absent the photon's quantum of light energy and and hence, absent its corresponding mass, (if that removal were possible).

Q: How are fabrons physically different from warpons?

A: Fabrons probably occur in clumps (or singularities) that are relatively fixed in space while expanding (with) it. Fabrons can be thought of as units of energy or "particles" that form the space network. Nevertheless, their flexibility and expansiveness sometimes makes them seem mobile. In fact, it is probably (globally) unified motion of fabrons that is referred to as expansion of space.

Note: The ability of fabrons to transport massive objects through space may actually be compensation (in reverse) for their inability to transport themselves through space. Since the fabrons form the space network and are thus unable to move, then any massive object it encounters is able to be moved … and (with sufficient transport) usually is.

Warpons are singletons (not singularities) and, when not constrained, have the ability to move through space together with their associated massive object whenever and wherever it moves. Warpons are probably attached to their object or incorporated into its matter.

In fact, what we call the "mass" of an object is simply a measure of the quantity of warpons that travel with the object. Otherwise, the primary differences between fabrons and warpons lie in their function.

Since all warpons of an object are attached to the object, it could generally be imagined as though the warpons are attached to each other, forming the "surfboard" with which the object "glides" through the fabron waves.

Locally, fabrons all act independently as particles of the expanding network while matter, photons, and other particles travel through it.

Mass vs. Energy

The motrons of a warp field form a protuberance (of warpons) within the space network. This protuberance then provides an underwarp to all components of the massive object within its subfield. It also provides an overwarp to all other massive objects in the superfield beyond the massive object.

The "surplus" motron singularities that are not associated with any warp field provide an energy field of fabrons that form the familiar space network. This underscores the similarity of mass and energy.

"***Warp Concept***" will posit warpons as the source of mass in matter and in dark matter. One potential source of a possible under-count of mass is the dark matter of the supposedly massless bosons. This is especially true of photons. (Ponder the irony of shedding light on photons!)

However, bosons are thought to generate "equivalent mass" that for some of the reasons mentioned earlier is not currently considered as mass until it is needed to complete an inventory.

Warpons probably constitute the mass of all material objects. The same warpons also likely provide the mass for themselves as the "accompanying" dark matter. If the warpons were counted twice, this might seem to imply an over count in the total mass of the universe. That total mass is already substantial.

However, since warpons are responsible for the mass of all massive objects, then much (if not all) of the mass of warpons has already been accounted for. The same warpons cause both inertia and momentum and are also responsible for the gravity (and weight) of the object.

"Space Warp"

"***Warp Concept***" posits that a dark force on object A (from the pressurewarp associated with some nearby object B) need not (and should not) be thought of as a distortion (or warp) of space itself, but does focus the warp field of A. This focus is primarily on one side (the far side) of a warp field effectively compressing (ruffling) the warp on that side and has the same effect as an afterwarp on that side. This will be discussed below in "***Gravitation Concept***."

This distinction between pressurewarp and afterwarp makes little difference in performance. However, it greatly simplifies the understanding. This example might also help illuminate another distinction previously made.

The warp field of object B provides its own internal warp called an "underwarp." It also provides an external warp called an "overwarp" to the remainder of its superfield. It is important to understand that overwarp and underwarp are two functions of the <u>same</u> warp field. The external warp provided by B will be called a "pressurewarp" on A. The effect of B's pressurewarp on A is called "gravitation." Stay tuned. Details and *narrative* to follow in "***Gravitation Concept***."

An overwarp on object (A) caused by the warp field of a nearby object (B) is called a "dark force." The effect it has on the warp field of object A clandestinely appears as a percentage increase in afterwarp depending on object A's own internal warp and to a lesser extent depending on any motion of object B. This is likely due to the distortion of object B's warp field caused by its own motion.

The dark force of object B provides a seemingly unbalanced pressure to the warp field of object A. There is a balance however. It comes from a consideration of the other side where the warp field of object A provides an equal and opposite dark pressure on the warp field of object B.

This so-called gravitational "attraction" is actually caused by the two dark forces pressuring or pushing the objects together. Of course, this is just semantics. This pushing from behind can still be called "pulling" since the <u>source</u> of the push is primarily in front.

You would not call it "pushing" when people embrace loved ones. It is thus proper to say that two massive objects pull or attract each other (from their location in front) with their gravitation "arms" (from behind).

The source, structure, and properties of a warp field will be detailed in this discussion. Every massive object has its own warp field (of warpons that is somewhat similar in structure to the fabrons of the "fixed" space network on the sub-atomic level. However, a warp field …

(1) … is a "free-floating" independent subspace within the space network,

(2) … is physically attached to the object,

(3) … travels with the object wherever it goes and stays wherever it stays,

(4) … provides mass to the object,

(5) … has a fixed, inelastic density, and

(6) … generally exist in singletons as opposed to the singularities of fabrons.

For these reasons, motion can be thought of in terms of the relationship between the dark energy singularities of the space network and the dark matter singletons of individual warp fields.

It would be difficult to overlook the similarity of the concept of afterwarp to the concept of "space warp" from general relativity. This similarity partially motivated the naming of "warp" fields. "***Mass Concept***" now recommends that afterwarp become the new space warp. Before this can be done, however, some of the differences should be noted. "Space warp" of general relativity supposedly displays the following properties.

Properties of "Space Warp"

- Space, itself, must be curved rather than just the energy of the local warp field. **MU** maintains its superior simplicity by allowing (and providing) an entire universe that is Euclidean. (This is an important consideration since no theory of geometry easily accommodates (or provides a *narrative* for) space warps of (or within) Euclidean space or even of (or within) non-Euclidean spaces.)

- Orbital paths must be controlled indirectly through complex space warps caused by gravitation rather than directly by gravitation as in **MU**.

- Besides increased complexity with space warp, the only difference compared to afterwarp (with gravitational orbits) seems to be that photons would not be required to have even minimum (quantum) mass as is required to support motion in **MU**.

- There is no *narrative* for space warp. Calculations can be made using the derived formulas based on the observed behavior assumed to be caused by space warps.

- However, it is unknown exactly how matter could physically achieve the "warping" of space? Must we assume that warping under pressure is just the way space behaves?

- Due to gravitation equivalence (with space warp), motion of massive objects is shaped by their warp fields and gravity is also a function of warp, then motion of a massive objects should be detectible from a great distance as a fluctuation in its warp field. In today's lexicon, this fluctuation might be referred to as a "gravitation wave." More on this interpretation later.

Imports of "*Mass Concept.*"

Mass has always had two definitions, one formal and the other informal. *Formally*, mass is the proportionality "constant" between force and the resulting acceleration of any (specific) moving object. This "constant" becomes a variable when we consider a variety of objects in a variety of shapes and sizes.

Informally, the mass of an object is a measure of the quantity of matter that comprises that object. When the concept of mass is referenced in the study of motion, the two definitions are incompatible. This requires the invention of "mass equivalence" for reconciliation.

"***Mass Concept***" attempts to reconcile these two definitions into a usable understanding of the concept and the implications of the role of mass in the achievement and understanding of motion.

"***Gravitation Concept***" will continue this reconciliation and simplification. Neither of these concepts requires the existence of "equivalent mass" or "space warp." These "devices" will only be used to help explain the two *narratives* to those who understand mass and space in these terms.

Mind's eye "visualization" of fabrons and warpons is a useful way to study and communicate the behavior of various aspects of the universe. However, it is important to understand that the purpose of this "***Mass Concept***" is to help understand how it works, not necessarily how it looks.

It will be left for particle physicists to determine if the newly discovered Higgs boson is what we have been calling a "motron" in **MU**. If this is the case, then since there appears to be only one type of Higgs boson: then it is probably just the warpon. The existence of this particle was suggested by Peter Higgs and others in 1964.

Furthermore, the density of fabrons may be adjustable by "singularity reduction." Singularity reduction will be discussed in "***Expansion Concept***." The epic story of the physical universe could be called the "Drama of Motrons."

Conservation of Momentum

We have now established the narrative for all three versions of momentum. This leads to the three versions of conservation of momentum. These can be used to introduce the corresponding three concepts of gravity that will be discussed in "***Gravitation Concept***" below. In this book, we will concentrate on the linear version of gravity. This introduction requires specific notation such as $f_e = m_0 \cdot a$.

The "God Particle"

We can be confident that motrons are not "God particles" even in jest. If "***Mass Concept***" is correct, then even monopons have (quantum) mass. "Q-mass" will be discussed in ***Advanced Speculation***. If all particles (except fabrons of the space network) have mass or at least q-mass, then there is no need for a "God particle" to assign mass. In a less majestic fashion, it has already been shown how the warpon "determines" as well as "provides" the mass of all massive particles.

Anyone searching for "answers" in this regard is looking in the right place, but with the wrong tools. This search for "the answer" is a search for "why." We are told that science cannot tell us why. However, why not?

Things like "beauty" are decisions, not truth. No aspect of architecture, for example, should be excluded from study just because some may use it to support their concept of beauty. Nothing should be included for that reason either. The order in the universe implies that everything has a reason, which we are calling a *narrative*. Narratives should not be included in, or excluded from physics just because some may use them to illustrate order in the universe.

Thus, to find answers, it is first necessary to stop "not looking." Blinders are not tools of science. Nor is a muzzle. *Narrative* physics finally provides the tools to search for reasons. (For more on monopons, please continue to "***Gravitation Concept***" next.)

9. Gravitation Concept

Simulated Afterwarp

As discussed in "***Ruffle Concept***," there is another way, besides a general force, to cause unbalancing of the warpons. Here is an example. Suppose object B drifts into a significantly dense portion of the warp field of a much larger (and heavier) object A.

The long dense warp rays of A, which will be labeled (A), interact with proportionally long dense warp rays (B). As mentioned above in "***Mass Concept***," pressurewarp (A) does not directly affect B's motion, but it "compresses" the like-directed warp on the far side of B inward toward object B in a way very similar to the effect of B's own afterwarp (B) when traveling toward A.

Simulated Revving

This simulated afterwarp, "ruffling-without-revving" of (B) by warp (A) on the far side of B, behaves as if it were caused by revving of B, ruffling afterwarp (B). Thus, B's own ruffled afterwarp (B) causes B to begin moving toward A. This ruffling of B causes "real" afterwarp that adds to (B). Pressurewarp (A) continues to simulate afterwarp (B), which increases ruffle-causing motion, which further increase afterwarp (B), which produces greater acceleration of B. The complex process described in this paragraph will be called "simulated revving."

Meanwhile, (B) is analogously performing simulated revving of like-directed warp (A) on the far side of A. However, (A) cannot perform simulated revving on the near side of (B) nor (B) on the near side of (A). The two pressurewarps cannot rev oppositely directed warps between the objects. In general, warp fields cannot pull; only push.

Using Calculus

Anyone who insists on trying to understand physics mathematically should compute the one-body momentum and apply that result to the two body problem by integrating over the warp field of each body separately. This will not be pursued here.

Monopon Gravitation

Notice that in order for gravitation to increase the speed of any object it must increase its afterwarp. Since gravitation cannot increase the afterwarp of any monopon, then neither can it accelerate a monopon.

This narrative confirms what we already knew about universal c being unaffected by gravitation. This agreement mutually tends to support the above narrative of gravitation.

Pressurewarp

Q: Suppose the pressurewarp of some massive object A is "squeezing" the warp field of a similar object B, without directly adding to the warp. How does that pressurewarp produce gravitation?

A: The complex technical answer involves the angles and the components and intensities of the warps. It is best formulated using vector terminology, which tends to obscure the narrative.

Water Experiment – Part 1

To ***understand*** the answer, consider this experiment. A container of water is being weighed on a scale in front of you. Reach out your hand and touch the water without touching the container or the scale. What causes the scale to register an increase in weight? After all, your hand is still attached to you and supported by you. It is not floating in the water! (Queue organ music.)

The more of your hand you insert into the water the greater the indicated weight gain. By studying the situation, you can discover empirically that the scale is not adding the weight of your hand, but the weight of the water displaced by your hand.

When you insert your hand into the water, you can feel the (buoyancy) force required. You can tell that you are pushing the water down when you notice the water level rising around your hand. The force you are exerting on the water is being transferred to the scale. If the rising water is diverted or allowed to overflow away from the scale, then the scale does not register any further "weight gain."

Analogously, the effect of the pressurewarp of A upon the far-side warpons of B was transferred to object B causing motion of B. This motion ruffles the afterwarp on the far side of B and thus causes acceleration of B toward A.

This increased afterwarp continues on both objects. Acceleration is caused by force. This theoretical force has a remarkable similarity to gravitation. "***Gravitation Concept***" will assume that we have finally found the source of gravitation. If "***Gravitation Concept***" can be established, then the "pressurewarp" cause of gravitation will finally be understood. However, we are not done yet.

Near Side Effect

At least one detail that has been temporarily oversimplified should now be addressed. To complete this narrative, it is necessary for the warpon's effect ***between*** the massive objects to be considered as judiciously as that on the outside.

We have seen how the warp of massive object #2 can simulate the like-directed warp of object #1 creating simulated afterwarp of object #1. This will result in "partial simulated cancellation" of some of the afterwarp of object #2. This transfer of warp is called contra-warp.

The simulated cancellation that generally occurs between two massive objects reduces the repelling forces that would otherwise be pushing the objects apart.

In other words, the oppositely directed warpons between the objects would normally be "pressuring" them apart. Because of the simulated cancellation, however, this affect is reduced. Any loss of repulsion in a balanced interaction can be considered an attraction.

Water Experiment – Part 2

This thought experiment might help to understand how it works. In "Part 1" of the water experiment, pushing down against the buoyancy force of the water increased the net weight of the container on the scale. However, the water buoyancy must also have decreased your weight by an equal and opposite amount.

Try the experiment again with you and the water on separate scales. This will verify that the weight loss on the near side of the experiment should be equal to the weight gain on the far side.

Since each object is concentrating its external warp on the far side of the other, then its internal warp on the (common) near side must be diminished by an equal and opposite amount. This far side increase in warp will be called, "warp-gain."

Accordingly, the nearside pressure decrease will be called, “warp-loss.” Combining these two ideas, we can say that the pressure was effectively transferred from the front to the back. This interchange of warp will be called, “contra-warp.”

Narrative for Contra-warp

If two large objects (not shown) are mounted vertically (one above the other) by springs with the top one also held up by springs, it may be possible by standing between the objects to push the bottom one down with your weight. If you then try, while standing on the bottom one, to pull the top one down with your weight, the transferred weight with which you are “enhancing” (increasing) the top object will now be “diluted” (subtracted) from the bottom one.

The same weight cannot hold down both objects. It can only hold down one or a fraction of each. A careful analysis of contra-warp reveals that outer attractive forces of both objects are not diminished by the proximity of the other object. However, all inner repulsive forces on each object also supply contra-warp to the outer object. Thus, it is impossible to increase one without decreasing the other.

Figure 4 is a simplified diagram, representing (one side of) a gravitational attraction between two similar objects such as object #1 and object #2 shown here with typical afterwarp vectors (A and B) for each.

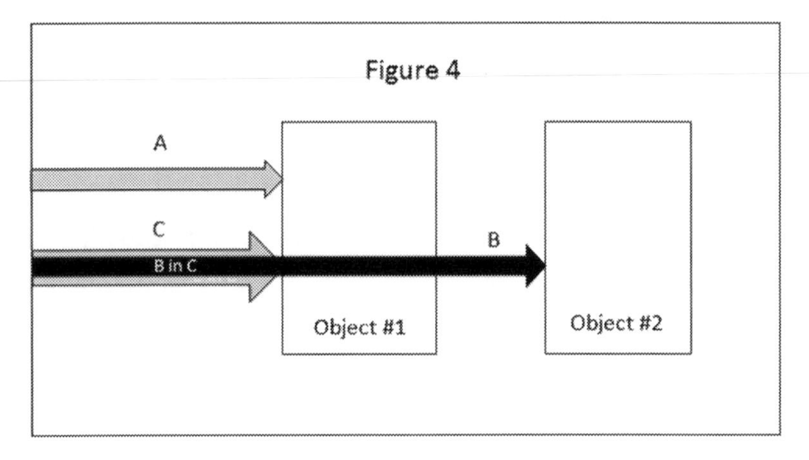

One-sided Gravitation between Object #1 and Object #2
(With Attracting-warp A, Repelling-warp B, and Contra-warp C.)

In figure 4, there are three types of afterwarp. We will discuss contra-warp C shortly. The other two are represented by afterwarp A of object #1 and afterwarp B of object #2. Since afterwarp A is pushing object #1 toward object #2, A is considered an *attractive* force. Contrarily, since afterwarp B is pushing object #2 away from object #1, B is considered a *repelling* force.

Most attractive forces (like A and C of object #1 in figure 4) are on the *outside* of the objects and are thus not diminished by the presence of object #2.

Most repelling forces (like B of object #2 in figure 4) are partially *between* the objects and pushing the objects apart.

In this latter case, afterwarp B of object #2 is somewhat between objects #1 and #2 and is pushing object #2 away from object #1. This has caused afterwarp B to become partially intercepted by object #1 creating an enhanced contra-warp C to join afterwarp A as attractive forces on object #1.

Note that although warps A, B, and C were originally equivalent, that contra-warp has enhanced C at the expense of B because B (of Object #2) must push through Object #1, which "filters" out much of its warp.

Three similar afterwarp and contra-warp forces (not shown) are pushing in the opposite direction on both objects. The net result of all contra-warp forces is increased attraction and decreased repulsion. "When two massive objects are sufficiently close, they both experience equal but opposite contra-warp."

There is one additional observation to be made. Normally, afterwarp A and afterwarp C would not appear in the same illustration. While afterwarp A would be typical ***before*** the presence of object #2. Enhanced afterwarp C ***requires*** the presence of object #2.

In summary, the increasing afterwarps (due to decreasing distance) of both objects, as they accelerate (due to gravity) toward each other (not shown), have caused a partial cancellation of oppositely pushing forces from the forefronts of both objects. This has resulted in increased contra-warp provided by two objects being attracted (pushed together from both sides) but with diminished repulsion from the middle.

These two (complete) forces add together in a natural way to produce the "force of gravity" pushing both objects together.

Note: this is not a ***single*** force as would occur in a tug-of-war with just one rope. In this case, each object is attracting the other object. Thus, the first object could attract the second even if the first were constrained.

This *narrative* would be simpler (but it would require somewhat more verbiage) if the two objects were of substantially different masses, but the concept is the same. This *narrative #16* completes the concept of gravitation and finally removes "gravitation" from the realm of regality.

This might be a good time to mention some of the variety in the types of regality that dwells in that ill-fated realm.

- Some common regal concepts are those that we understand what they do, but not how they work.
- A second level of regality occurs when concepts we basically "understand," seem to have a few "mysterious" ramifications.
- A third (less malevolent) level of regality occurs when a concept is assumed to be totally incomprehensible.

Newton's Third Law of Motion

Before leaving this topic, one additional consideration might help with the visualization. Even when we remember that the range of gravity is very great, it still *seems* unlikely that a smaller than 8 thousand mile diameter object (like the earth) can have any effect on the far side of the over 800 thousand mile diameter object (like the sun) that is 93 million miles away. Just remember that any huge object is just a huge accumulation of many small components.

It might be easier to visualize the pressurewarp of the small object on some of the closer components of the large object. The effect on the large object is the sum of the effects on all of its components.

The relative sizes and masses of the two objects, of course, are immaterial to the fact of gravitation. Yet, mass does affect the value of relative motions. By Newton's 3rd law, the force is the same value, but oppositely directed, on both objects.

By Newton's 2nd law, though, the less massive object will accelerate more than the more massive one. In fact, any time two objects affect one another, the corresponding forces will be equal and opposite.

In other words, whatever forces the external warp of object A applies to object B, object B applies the same but opposite forces (reactions) to object A. This is Newton's 3rd law of motion.

The Equivalence Principles

There seems to be little difference between the linear acceleration due to gravitation caused by pressurewarp and the linear acceleration causing by some manual increase in afterwarp.

Einstein reasoned that, within a closed room, there is no way to tell the difference between the perception of the force of a gravitational frame and that of a manually accelerated frame. This was discussed in "***Hyperspace Concept***" earlier.

This seems to be true since field size is neither visible nor measurable directly. We now know that both experiences are caused by changes in the invisible afterwarp.

The only difference is whether the afterwarp increased a) *naturally* due to the presence of a nearby pressurewarp indirectly causing the physical increase or b) *materially* due to a manual motion of the object directly increasing the afterwarp.

Understandings *for* a) and b) are:

a) The warp field of nearby object A simulates afterwarp of B causing linear motion of B. This increases the physical afterwarp on B causing acceleration of B.

b) A manual force applied to B causes the same motion of B. This similarly increases the afterwarp of B causing the same acceleration of B.

From inside of closed, accelerating container B, there would clearly be no way to tell if the acceleration being felt was caused by the gravitational force of a nearby afterwarp field as in a) or some manual (or space warp) force as in b). "***Gravitation Concept***" has removed all mystery from this equivalence principle.

Afterwarp and pressurewarp both cause acceleration by revving. This is done by increasing warp concentration on one side of a massive object at the expense of the other.

Thus, it is not surprising that the following two types of force are equivalent. (Gravitation equivalence was referenced when discussing the cause of gravitation waves in "***Gravitation Concept***" earlier.)

a. A gravitational force on an afterwarp field, versus

b. A manual force causing, and then continually increasing the waxing afterwarp field.

This relation will be called, ***"afterwarp equivalence."*** This *narrative #17* now removes Einstein's "equivalence principle" from the realm of regality.

It is interesting to note that, *lacking a space network*, a gravitational force would seem to be some kind of autonomous "gravitation wave." Until now, physicists have been forced to assume this autonomy.

The Second (new) Equivalence Principles

There would seem to be little noticeable difference, from inside a closed, rotating chamber, between precession caused by anglewarp pressure and the precession caused by manual pressure.

The only difference is whether the pressure occurred a) *naturally* due to the presence of a nearby anglewarp indirectly causing the physical increase or b) *physically* due to a manual motion of the object directly increasing the anglewarp.

It turns out that not only can pressurewarp be simulated with manual force; the same is true of anglewarp.

a) The warp field of nearby object A can cause an anglewarp on the axis of rotating object B. This can introduce re-orientation of B's axis, causing its precession.

b) Manual force applied to the axis supports of B can introduce the same re-orientation of B's axis causing the same precession.

From inside of closed, rotating chamber B, there would clearly be no way to tell if the precession being felt was caused by the precessional force of a nearby anglewarp field as in a) or some manual (or space warp) force as in b).

"***Gravitation Concept***" has now removed all mystery from this second equivalence principle.

Thus, it is not surprising that the two effects are equivalent.

a. ***Precession caused by an anglewarp field.***

b. ***Precession caused by a manual force.***

This relation can be called, ***"anglewarp equivalence."***

It seems likely, given time, that someday "precession waves" will be measured.

Revolutionary Motion

If afterwarp provides linear motion and spinwarp causes rotational motion, then what causes revolutionary motion? The motion of revolution seems to be a cross between linear and rotational motion. This type of warp might be called, "orbitwarp."

Moreover, the expression "orbitwarp" seems somewhat closer to being afterwarp that it does to being anglewarp. Thus, it might be easier to modify our notion of afterwarp to accommodate motion of revolution. It only seems necessary to add either a physical or gravitational, or possibly a magnetic "radial arm," to an afterwarp to establish orbital motion.

Orbitwarp is clearly not a new type of warp since removal (or lengthening to infinity) of the radial arm causes the motion to revert to linear. It seems prudent just to consider orbitwarp to be a special version of afterwarp or a cross between afterwarp and anglewarp.

Nielucion

It now seems likely, and "***Mass Concept***" will assume that the "advancement" of Mercury's orbit (discussed in "***Angular Momentum Concept***") previously thought caused by "space warp" is *narratively* caused by the (circular) spinwarp (field) of solar rotation.

It appears that the spinwarp of object A's rotation (the sun in this example) can have an acceleration (direction, not speed) effect on the orbit of object B (Mercury in this example).

This effect is similar to the acceleration effect on the motion of object B that is caused by the gravitational force of object A's afterwarp.

This new (orbitar) version of precession that results from orbitwarp will be given the new name "*nielucion.*" This *narrative #18* now removes "nielucion" from the realm of regality.

The Third Equivalence Principles

It would theoretically be impossible from within an affected container to distinguish the orbitar acceleration of a nielucion field from that of an externally rotated anglewarp field.

This explains how, lacking a *narrative* for "orbitwarp," a nielucion field could have been thought a "space warp" by some earlier pioneers of physics.

Stated differently:

a) The spinwarp field of nearby object A can induce orbitwarp on a possibly eccentric orbit of B around A. This can cause nielucion of B's orbit around A.

b) A manual spin force applied to the orbit of B around A can induce the same orbitwarp of B's orbit that produces the same nielucion of B's orbit around A.

From inside of closed, container B, there would clearly be no way to tell if the nielucion of B's orbit around A was caused by the effect of A's spinwarp on B's orbit as in a) or by some manual (or "space warp") force as in b). "***Gravitation Concept***" has removed all mystery from this third equivalence principle and from the notion of "space warp".

Thus, it is not surprising that the two effects are equivalent.

a. Nielucion caused by an orbitwarp field.

b. Nielucion by a manual force (such as a supposed "space warp").

This relation can be called, ***"orbitwarp equivalence."***

It seems likely, given time, that someday even "nielucion waves" will be measured

Since nielucion was unrecognized prior to **MU**, it is not surprising that nielucion is currently referred to as "precession." It is easy, now, to recognize the difference between precession and nielucion.

- *In precession*, an anglewarp causes "rotation" of an axis around the center of rotation.
- While, *in nielucion*, an orbitwarp causes "rotation" of an entire orbit around the orbital center.

We will discuss other similarities between precession and nielucion just below. The similarities of these two with gravitation will also be presented in Table 4. This table suggests that "orbitar precession" may be yet another name for "orbitar nielucion."

Nielucion as Space Warp

Anyone who wished to continue thinking of nielucion as a space warp is clearly justified in doing so. However, for consistency, they should also consider gravitation and precession in those terms. Lacking a narrative, though, the details and justification for that "warp-view" of space will not be provided in this book.

Reactions to Warp Forces

Just as it is more difficult (than the converse) to measure the effect of Earth's gravity on the motion of the sun, it is even more difficult to measure the effect of Mercury's orbitar pressurewarp (nielucion) on the rotation of the sun. This effect can be expected and probably calculated, and will likely someday be measured.

If this eventual measurement were found comparable to the observed nielucion (precession) of Mercury's orbit then this would confirm the concept of nielucion.

One word of caution, however. The rotation of the sun is causing orbitar nielucion of not only Mercury's orbit, but also that of the entire massive content of the solar system.

Properties of Nielucion

Some might wonder why precession and nielucion seem to require objects to be rotating or revolving whereas gravitation can occur between stationary objects. This is a complex question with a very simple answer.

Precession and nielucion both require warp fields like gravitation does, but they also require turn centers for their anglewarp. Gravitation only requires a linear warp field (afterwarp) so no turn center is necessary. In any force field situation where there was no turn center, then we would just call it gravitation.

We know that the warp fields of two massive objects can produce an appropriate anglewarp if one of the object is rotating and the other is revolving around it (especially in the same direction).

It seems likely that retrograde (oppositely directed) nielucion would also be possible, but this would likely occur naturally only rarely. In addition, the interaction caused by retrograde nielucion would be expected to cause degradation of the motion, which would tend to shorten the lifespan of the orbit.

Another opportunity for nielucion might occur if two adjacent objects were both rotating, especially if rotating in opposite directions. Another possibility might occur if a dense fluid were streaming past a rotating object, even if no physical contact were made.

It should also be noted that examples of nielucion are probably much more common than might now be suspected. Nielucion is more obvious when the orbit of one around the other has noticeable eccentricity.

Nielucion does not (usually) distort the shape of an orbit; it simply revolves the orbit (including its eccentricity) around the center of gravity of (the rotating) object A. This might be called orbitar precession.

This "advancing" of the orbit is common; it seemingly shortens the period of revolution if this "sidereal" orbit is considered relative to the stationary background. We could call it, "sidereal orbit advance."

Without compensation, nielucion could increases the calculated mass. (E.g., if this seemingly shortened period is not taken into account in every such orbit, it could slightly overstate the calculated mass of one or both objects accordingly.) This understanding of celestial mechanics might necessitate slight changes in some calculations or in the interpretation of some results.

Not knowing the source of gravitation for all these years has prevented the recognition of the similarity between the source of gravitation and those of precession and nielucion.

Due to this similarity, it may turn out that the "linear precession" being sought is actually a new perspective on gravitation (we previously noted the similarity in "***Angular Momentum Concept***").

These categories are not a matter of substance; they are simply a matter of classifications or "optional tagging." Since they all cause acceleration, then they can all be classified as "forces." If these classifications are valid, then the three names for these three warp forces should probably be **"gravitation"** that is short for *"linear gravitation,"* **"precession"** short for *"angular precession,"* and *"orbitar nielucion"* will be called **"nielucion."**

These three forces could be called the "warp forces." With these classifications, then the name "gravity assist" should remain appropriate and unchanged. To sort this all out, it might be time for another chart. See Table 4. If this analysis is correct, then the three names in each column are three possible names for basically the same thing.

Warp Forces Table 4	**Linear**	**Angular**	**Orbitar**
Common	***Gravitation***	***Precession***	***Nielucion***
Gravitation	L. Gravitation	A. Gravitation	O. Gravitation
Precession	L. Precession	A. Precession	O. Precession
Nielucion	L. Nielucion	A. Nielucion	O. Nielucion

One possible alternative to the above analysis is the chance that the three versions of each warp force need not necessarily be identical. In that case, there might be as many as nine (or more) different warp forces. For now, we have enough to consider with just these three.

Monopon Warp

A single photon of light probably has one warpon. This type of particle is being called a "monopon." Obviously, a "monopon warp field" of one is always: "unbalanced," "nonsymmetrical," "displaced," and "distorted." Since the field has maximum afterwarp of 1, the photon will travel with maximum speed with respect to the space network. (Expect more on "constant frame" in "***Relativity Concept***" next.)

There are at least six major monopon implications:

1. Any monopon particle cannot physically speed up since there is no physical way to create afterwarp greater than 1 by shifting its warpon around. This is one of the main reasons photons cannot travel faster than parochial c.

2. No monopon particle can lose speed or remain motionless simply because there is no way to reduce or balance the

afterwarp. This is the main reason that (in a vacuum) photons never travel slower than parochial c.

3. When a photon approaches a massive object, its warp field already has maximum afterwarp. Therefore, it cannot be speed-accelerated by the massive object. This implies, by our new understanding of Newton's third law that a photon or many photons cannot accelerate any massive object being approached. This is one of the main reason the monopon is thought of as being massless.

This does not prevent a massive object from "warping" or curving the trek of a photon as opposed to actual warping (or bending) of space. (Any change in quantity or direction of speed is considered to be acceleration.)

4. Thus, since m = f/a and a gravitational force normally causes a stream of photons to accelerate (not speed but direction) inward, then a photon clearly has "real" mass that we will call q-mass, not just "equivalent" mass of its light quantum.

5. One warpon each might be all the gravitation that would be necessary to link a string of photons to form a light ray. This *narrative* might explain the ability of photons to travel as a particle-wave.

6. If this structure is correct, then even though light rays can be affected (bent) by gravitation, it is not likely that they can be affected by other light rays. This relation deserves further analysis. The effect of a gravitational field on a light ray also deserves further analysis. (Whoever thought that starlight … the energy, like starlight … the idea, could be so weighty?)

A light ray is probably polarized by its "quantum gravitation" (gravitation associated with the q-mass of a light quantum), but a light beam could be a collection of light rays with random electric field orientations. The idea that photons have mass would explain why black holes are dark. Photon mass would cause light rays to be bent back by extreme gravitation. If light rays can be bent by gravitation, then it would not be surprising if photons of every frequency and other massive particles such as warpons and even neutrinos could be found in orbit around super massive stars such as black holes. This would make it like a giant cyclotron or super collider.

This speculation raises many interesting possibilities and further study will most likely provide answers and solutions to numerous particle-related quandaries and more. However, this aside will not be pursued further in this book.

Imports of "*Gravitation Concept*"

"Physics has come a long way in the 300+ years since Newton declared gravity to be a natural force. The failure to find the source of this natural force or the nature of this force of nature is not for lack of trying. The search goes on even today. Does the Higgs boson ring any bells? What about the search for gravitation waves? If the source of gravity revealed in this document is certified, then this one discovery will simplify and motivate much new searching and researching by many new searchers and researchers."

Fundamental Force

All acceleration is caused by force; the warp forces are: (linear) warp, angular warp, and orbitar warp. They are called respectively, gravitation, precession, and nielucion.

Therefore, warpons are the elementary particles associates with the weakest but most far-reaching of the four fundamental forces. It seems appropriate to call these forces the "warp forces." Any force similar to a warp force (except that the corresponding force field is replaced with a "manual" force) will be called a "manual-warp force."

Since a warp force is first-of-all a force, then the symbol for warp forces could probably continue as "F." When more specificity is called for, a subscript could be added, such as "F_g" for gravitation, "F_p" for precession, or "F_n" for nielucion. If we ever need a generic symbol more specific than F for "warp force," then "F_w" would probably suffice.

The Major Warp Force Classifications

There are currently three major warp force classifications. In the future, there may be other warp force classifications or sub-classifications. The current major warp forces will be numbered and summarized next.

1. Warp force #1 is gravitation, F_g. It can be described as an afterwarp distortion of another afterwarp.

2. Warp force #2 is precession, F_p. It can be described as an anglewarp distortion of a spinwarp.

3. Warp force #3 is nielucion, F_n. It can be described as a spinwarp distortion of an orbitwarp.

Cross-Classifications of Warp

Warp Unification Table 5	**Primary**	**Secondary**
Direct Cause	Linear Motion Angular Motion Orbitar Motion (Afterwarp/Anglewarp)	Linear Reaction Angular Reaction Orbitar Reaction (Afterwarp/Anglewarp)
Indirect Effect	Linear Momentum Angular Momentum Orbitar Momentum (Vectorwarp)	Linear Gravitation Angular Precession Orbitar Nielucion (Pressurewarp)

Warpon Waves

From what we have learned here about warp and from what we discussed earlier about long predicted gravitation waves, it seems likely that we should someday discover other warpon waves in addition to gravitation waves. This should include "precession waves" and "nielucion waves." It does not now seem likely, but might there be others such as various momentum waves?" However, these will be very weak and in other ways also very similar to gravitation waves. It might initially be very difficult to distinguish between any of them. It would be earth shaking (almost) just to detect them. Would it not be a good idea for a generic name for all of these waves? Since these "waves" are more like "ripples," I might recommend, "Warp ripples." On the other hand, does that sound too much like ice cream?

9th Uni-Verse

"In order for gravitation
To increase the speed of any object,
It must also increase its afterwarp."

10. Relativity Concept

Initial Assumptions

There are twelve assumptions we will call the "Extended Relativity Conjecture," that virtually all physicists make that allow them to accept The Special Theory of Relativity. This is acceptable in spite of its very high Occam simply because it works … not because it works simply.

1. The universe is homogeneous,
2. The universe is isotropic,
3. The universe is infinite,
4. The universe is expanding,
5. Time is the 4th dimension of space-time.
6. All motion is relative (this is called "classical relativity"), and
7. Both the "parochial" and the "alien" speed of light in a vacuum are constant. (This constant speed is called universal c.)

"Epicycles" of Special Relativity

The problem in this relativity conjecture is not error or inconsistency, but complexity. The evidence of complexity is not just the fact that there are seven assumptions above, but also by assumption "creep" and by assumption "contrivance."

The ancient "Copernican principle" contemplates: "Earth does not occupy a position of any significance in the universe."

1 & 2 together are called the "cosmological principle." This principle is a modern attempt to codify the Copernican principle and carry it to the next logical step. This rephrasing of the Copernican principle, though, is somewhat more from the perspective of the universe in space-time, than it is from Earth in the universe.

Assumption 3 is necessary for assumptions 1 & 2 to be possible. Assumptions 1 – 3 require the addition of assumptions 4 and 5 (as they are now crafted). **

Assumptions 1 – 5 do not necessitate assumptions 6 or 7. Assumption 6 is independent, as is parochial c of assumption 7. However, *if* classical relativity of assumption 6 and alien c of assumption 7 are both assumed, then this creates three additional requirements 8 - 10.

As a direct result of this, any of the three *contrived* requirements 8 - 10 can be established, but only if the (similarly contrived) alien c of assumption 7 is first assumed true. The "contrivance" cited above will be explained and illustrated in this "***Relativity Concept***."

The assumptions for The Special Theory of Relativity continue.

8. Time dilates in moving reference frames,
9. Length contracts in the direction of motion, and
10. Synchronicity is lost for moving observers.

Assumptions 8 – 10 are especially noticeable for rapid motion. The four assumptions 1 - 4 create two additional requirements within general relativity:

11. As speed v increases, mass m increases by the formula $m = m_0 \cdot (1 - (v/c)^2)^{-1/2}$, where m_0 is its rest mass **

12. Space becomes warped by gravitation.

(It acquires a non-Euclidean subspace.)

Beyond the 12 assumptions comes one basic requirement. Since all motion is relative, then nothing is ever known to be stationary. By assuming, as investigators, that *we* are stationary and all of *our* measuring devices are stationary, then everything of importance is in the same (local) inertial frame. This might substitute for the definition of a "parochial system." Parochial systems will be considered soon.

This substitution might be expressed as a rephrase of #6 to read, "All motion is relative to observers who must consider themselves and their instruments stationary for the purpose of their calculations."

** *Assumption #4 and #5, along with #11 and #12 are prerequisites for general relativity. Also, note that "rest mass" is determined relative to the "currently" presumed stationary reference frame.*

Meaning of "Constant c"

Suppose you were traveling through space at the incredible speed of 20% of c when you turned on your headlights. How fast would the headlight photons then be traveling (forward) away from you?

Believe it or not, there seem to be at least two correct answers to this question.

1. The photons are racing ahead of you with an alien rate of progress $S_A = c^2 - v^2 = (100\% - 20\%) \cdot c = 80\%$ of c.

2. The photons are racing ahead of you at 100% of speed c.

The two answers are each factually correct depending on your interpretation of the facts (your reference frame).

a. Answer #1 assumes universal time t. (Natural) Relativity tells us that the photons are still traveling at parochial c. Since your speed is only 20% of c, then the photons will be pulling away from you with an **alien speed** (with respect to you) that is 80% of c.

b. Answer #2 assumes universal light speed c so that the light speed is always 100% of c. Notice how this answer ***initially*** seems simpler than answer a.

Special (Forced) Relativity tells us that, due to the circumstances, time has dilated, distances have shortened and synchronicity has been lost so that the photons traveling *at the **universal** speed of light* in this interpretation will have a rate of progress (with respect to you) at 80% of parochial speed. This is the same result as #1, but the "actual" speed (before the synchronicity adjustment) is 100% if c.

In natural relativity, the "rate of progress" or "ROP" will be called the "alien speed S_A (with respect to the moving point of origin)." Note that ROP should be measured as "distance traveled, d divided by time of travel t" rather than as a "percentage-of-total" due to the expansion of the universe.

Both of these interpretations are equivalent. The primary difference between #1 and #2 is the complexity of the understanding and the complexity of the computations in special relativity (next).

Comparing Rate of Progress (ROP) With Alien Speed (S_A)

Let us now calculate the ROP and compare it with the alien speed compared to rate $v_0 = d/t = 1$. Using the Lorentz Transformation, we can calculate the ROP for photons in Special Relativity.

$$ROP = [d \cdot (1\text{-}v^2/c^2)^{(1/2)}] / [t \cdot (1\text{-}v^2/c^2)^{-(1/2)}] \cdot c^2 = d/t \cdot (1\text{-}v^2/c^2) \cdot c^2$$

$$\mathbf{= d/t \cdot [c^2 - v^2] = c^2 - v^2 = S_A,}$$

with respect to a speed $v_0 = d/t = 1$, where d = distance, t = time.

This is the exact alien rate as that computed by natural relativity <u>without</u> using these conversion factors. (See answer #1 above.)

Keeping It Simple (Uncontrived)

Occam's razor generally distinguishes between two competing explanations of the same phenomenon based on which conjecture requires the fewest and simplest assumptions (and contrivances). Summarizing this idea, Einstein (and others) have said (in effect), "Everything should be made as simple as possible, but not simpler." [The occasion(s) and Einstein's exact words have apparently been lost; yet the sentiment is valid … and is often quoted.] This begs the question, "How do we know when our ideas are simple enough?"

To paraphrase another common saying, "Just because a demonstration makes something look simple does not mean it is."

When forced relativity was first published, the concept of "aether" (sometimes ambiguously spelled "ether") was just being repudiated based on numerous obscurities in its informal definition (there never has been an acceptable formal definition of aether) and based as well on a broad interpretation of the Michelson-Morley experiment. This will be discussed soon.

(Fabrons are different from aether primarily because fabrons are energy similar to photons and aether has generally been considered to be some type of elementary "vapor.")

Instead of fixing this, the idea was abandoned altogether. We must now wonder, "Why was the baby thrown out with the bath water?" The answer to this question is just a tiny bit obvious. The "random-events" bias ***against*** "narrative-physics" that led to this decision has temporarily sidetracked the subsequent development of physics. The discussion of this bias belongs more in the area of behavioral science instead of physical science. It will not be pursued further here. It is hoped that the redirection suggested in this book will eventually help reestablish the perception of order in the universe.

This unceremonious "loss" of aether left space with no fabric at the time. Photons were then expected to travel through space with no network. Imagining photons traveling with no means of support may seem simple at first.

Having no space network as an absolute reference frame made it easy to assume that, since starlight can be seen coming through space, then obviously "no space network must be needed." (This, of course, is another example of circular reasoning.)

Note: "*A more realistic possibility is that there really is a space network.*"

At the time, the simplest assumption seemed to be that "all motion is relative and light speed is a universal constant." Nothing else made sense at that time so this was as simple as seemed possible; therefore, it could not (and should not) be made simpler.

Imagining a space network was thought to be unwise if none was needed. (Please see Author's Note at end of "***Relativity Concept***.")

One question brings us back to earth. How does a photon "experience" the geometry of space if space is nothing but vacuum (which itself contains nothing but random energy). Just imagine the *narrative* for that! Also, imagine the properties that must be ascribed to that vacuum even to try making it all work. These were never done, of course, since narratives were not and (until now) are still not considered necessary.

The geometry of space must influence objects traveling through it; else what difference does geometry make? Why not make geometry "relative" and just use whatever works at the time. The obvious answer is that there must be a space network (originally called the "fabric of space"). Since the needed substance "should have been detectable" but was not detected, the required substance has become known as "dark energy." The properties that have been ascribed to dark energy are surprisingly accurate. Well done, ya'll!

The last time this happened and there was no space network to explain observed orbitar nielucion, it was decided that space, itself, must be "warped" to cause nielucion. Once "space warp" was credited with nielucion, it was also easy to credit it with gravitation. If not already, then "space warp" would soon have also been credited with precession. If anyone realized that we had no explanation for momentum, then "space warp" might have gained credit for that also.

To be more specific, "dark energy" is the name given by astrophysicists to the unknown substance that is exerting some kind of "repulsive force" on the universe causing "empty" space to expand. Strong evidence is presented throughout **MU** to support this "space network" version of "dark energy."

This validates (and is validated by) the space network of dark energy presented here. Fabrons have not previously been fully revealed in space because their nature and behavior have been unknown prior to this book.

Although other approximations and proportions have been made, the calculated density of energy in the space network is sometimes based on the relic afterglow of the Big Bang. This probably takes into account only that energy involved in the "bilarity epoch" and the subsequent epoch of inflation, which together initiated the Big Bang. "Bilarity" will be defined below within "***Advanced Speculation***." However, **MU** will assume that the (fabron) energy of the space network is distributed as singularities.

This is <u>not</u> an absolute requirement, but it might explain a few things that <u>are</u> going on. This will also be discussed in "***Advanced Speculation***."

With the above restrictions, it would be impossible for any stationary observer to use their stationary instruments to measure any distant motion while it remains transitory within any moving frame. Such "measurements" must be "calculations" that will be called "alien measures." This may seem easy to do. If however, the photons in the moving frame were to enter any measuring device within our stationary frame to be measured, its speed would necessarily adjust and the result could be (and nearly always has been) misinterpreted. This will be fortified with a *narratives* shortly.

Alien measures must therefore be calculations based on the 12 assumptions. Please note that calculations based on assumptions can rightly be considered assumptions as well. Because the "alien measure" of light speed is currently impossible to actually measure, then the alien measure of the speed of light has been *assumed* to have the same value as parochial c. This would totally make sense if there were no space network so that all motion was truly relative as once believed.

Contrary to popular belief, there are not "many" proofs, or even **any** *valid* proof and also no narrative for this assumption. This will be explicated (explained) shortly.

This alien measure of the speed of light will be called "alien c." Any calculations made by a stationary observer involving only events that are occurring within the observer's frame of reference will be called "parochial measures." "Parochial c" refers to the speed of light within the frame through which it is currently traveling or being measured. This will be called its "constant frame."

Since "parochial" c is the only value of c that can be measured, then this is all that ever has been measured directly. Thus, this is all that was shown to be constant (not just assumed to be) by the Michelson-Morley Experiment. There will be more on this below.

Since alien measurements are impossible to make directly, then the appropriate calculations will be considered "alien *measures*." This is common practice.

However, we must be careful not to use those calculations in any confirmation of the assumptions being made since that would produce a logical fallacy.

Assumptions (such as the assumption of constant c) that lead to conclusions (such as length contraction and time dilation) that are used in an attempt to prove the original assumptions to be true have also committed the fallacy of circular reasoning.

Comparisons of calculations with assumptions only serve to verify that the calculations were <u>done</u> correctly, not necessarily laudably or even appropriately or that the assumptions are physically valid.

For clarity, it is important to distinguish mental "measure<u>s</u>" based upon assumptions, from physical measure<u>ments</u>. Any alien measures and other mental determinations based upon assumptions should always be called "calculations" or "measures" to distinguish them from physical measurements.

Accuracy of calculations can be expressed by appropriate rounding of the result. Physical determinations can be called "measurements" or accompanied by the precision unit or the corresponding maximum error of the measurement. Calculated precision units of assumed measures are also assumptions (measures).

Equivalent Mass Problems

Allowing mass to increase by adding equivalent mass to a moving object creates at least two problems in forced relativity.

Equivalent Mass Problem 1

The relativity principal states that all motion is relative so that any object considered to be in motion must have the same mass (still be the same object) as when it is considered stationary.

Therefore, when a moving object is considered stationary, equivalent mass must inexplicably disappear just because we changed our coordinate system. This might make some mathematical sense if all "energy" were somehow equivalent to "some specific mass."

This is similar to the way the length of a moving meter stick becomes shorter than it was when it was <u>considered</u> stationary. (This manipulation of the truth makes far more sense mathematically than it does physically.)

Therefore, the <u>physical</u> mass of an object should be considered its (stationary) measured value called its "rest mass."

Excessive calculated mass is considered the "equivalent mass" of the energy associated with its motion. Therefore, to satisfy the mathematical assumptions of relativity, all observers making measurements (calculating measures) in forced relativity must consider themselves (and all of their instruments) stationary.

Equivalent Mass Problem 2

There are two versions of this second problem:

1. (Version i) For any $a > 0$, considering apparent mass above the base line ($m \geq f / a$) to be "equivalent mass" might exaggerate the quantity of matter thought to exist in the universe if matter is the only source of mass and not motion also.

2. (Version ii) Imagining that "equivalent mass" disappears when an object slows down might underestimate the quantity of matter thought to exist in the universe if matter is the only source of mass and not motion also.

Mass should not (physically) be considered a measure of matter if mass changes when its physical matter does not.

Natural Relativity

"Natural relativity," that follows is a "simplification up-date" of special relativity and general relativity combined ("forced relativity").

Because of its simplicity, natural relativity harbors no general necessity to distinguish between inertial frames as in special relativity and accelerating frames as in general relativity.

There are five basic assumptions in natural relativity. (Once understood they can be made shorter and more concise.)

1) The D3 space network of fabrons is always finite as it expands exponentially; (It is uniform with constant density D. This assertion will be justified soon.)

2) Warp fields provide acceleration proportional to the force causing the acceleration; mass and speed are independent variables, but $F = m \cdot a$.

3) All motion is relative to a specific frame; an object's current speed vectors (relative to serial frames) are additive as are all of its acceleration vectors.

4) Symmetrical warp fields correspond to stationary frames in which parochial motion is absolute.

5) Parochial light speed in a vacuum is constant c even as the universe expands, regardless of any motion of the common parochial frame of the source and the receiver.

These 5 assumptions are separate aspects of a single hypothesis called the "<u>the motion conjecture</u>."

Light speed can only be measured by instruments within its constant frame. This is not a new restriction since (by *narrative*) photons always adjust their speed to the parochial frame of the measuring instruments (#5 above). This is the main reason the Michelson-Morley experiment failed to detect any changes in light speed.

Clarification

Many assumptions of natural relativity are similar to the assumptions of forced relativity. Both systems assume,

1. Parochial light speed is always c,

2. Alien measures of the speed of light require that speed be calculated with respect to some (assumed) stationary frame of reference based on the assumptions of the asys. That measure is only ***valid*** within that system.

The initial difference between the assumptions of forced relativity and those of natural relativity is that forced relativity assumes constant alien light speed and variable measuring devices, whereas natural relativity (being presented now) assumes consistent device measurements and variable calculations of any alien speed of light. In natural relativity, only the coordinate system, and not necessarily the observers or their measuring devices, need to be assume stationary. It is only necessary that light speed be measured (or assumed) with respect to a parochial frame. It should also be pointed out that ***all natural relativity assumptions are supported by narratives***.

Two Linear Motion Problems in MU

There are two obvious problems in the analysis of linear motion within **MOTION UNIFICATION.**

Motion Problem 1

What happens when a force is applied to increase the speed of an object that is already in motion? Additional distortion of the afterwarp that is already displaced would not have the same effect as the original force. Worse yet, what happens when a force is applied to a warp field that is completely shifted so that the maximum afterwarp has already been achieved?

Motion Problem 2

The second problem has to do with special relativity. How do we explain the known effects of relativistic motion?

Solution to Both Motion Problems

As it turns out, in **MU** these two problems solve each other. Maximum speed would occur when maximum afterwarp was achieved (warp rate = 1). Thus, the maximum afterwarp would correspond to a speed of light speed c.

Any concern about the piling up of fabrons limiting mobility can be somewhat mollified by remembering that fabrons are energy and (since they have no mass) they do not "occupy" or "fill" space in the same way that matter does.

For a slowly moving object, acceleration is approximately proportional to the applied force according to Newton's 2nd law. The greater the speed of an object the greater is the required force above the base line (of m·a) to continue causing acceleration. This phenomenon will be called "base-line acceleration."

Base-line acceleration circumvents the need that seems to have motivated the invention of equivalent mass to support forced relativity. If mass m is to be the proportionality constant between force F and acceleration a, then moving objects oddly require more force to achieve the same acceleration as stationary objects because the mass of moving objects seem to "somehow" increases with speed. This supposed increase in mass in forced relativity is called equivalent mass.

However, the only reason the mass seems to increase is that mass has been defined as a proportionality constant. Like c in natural relativity, mass in forced relativity must change to remain constant. Contrarily, **MU** will retain constant mass of every massive object independent of changing speed.

The "Effective Force" Option

"***Motion Concept***" proposes that, rather than problematically adding equivalent mass to the mass of a moving object, that the definition of constant mass be faithfully restored to what is now called its rest mass m_0. When traveling at "relativistic" velocities, the effective force f_e could be increased (without changing the mass) from the old force f to the new value f_e by the FitzGerald "force-factor"

$$\gamma = \frac{1}{\sqrt{1-(v^2/c^2)}}.$$

Now we have: $F = f_e = f \cdot \gamma = f \cdot (1 - (v/c)^2)^{-1/2}$.

The value for the force required for "relativistic" acceleration would be the same as that calculated by the old method of adding $m_0 \cdot m_e = m_0 + \frac{m_0}{\sqrt{1-(v^2/c^2)}}$, where n is the speed of the object. This increased value for force in **MU** will be called "effective force" or f_e. So, by using effective force f_e instead of including "equivalent mass" that is supposedly added to the mass of the moving object by acceleration, the old formula ...

$$F = (m_0 + m_e) \cdot a = m_0 \cdot a + \frac{m_0 \cdot a}{\sqrt{1-\left(v^2/c^2\right)}} = (m_0 \cdot a)[1 + (1 - (v/c)^2)^{-1/2}]$$

changes to $F = f_e = (m_0 \cdot a)\,[1 + (1 - (v/c)^2)^{-1/2}]$.

The Baseline Effect

Clearly, this new process does not change the value of F from that obtained by using equivalent mass, but it does change the understanding considerably. The effect of the elliptical afterwarp on minimum force f will be called "the baseline effect." For "nonrelativistic" velocities, the force-factor γ rounds to 1 and can be ignored.

For "***Relativity Concept***," only the formula would change: Instead of increasing the mass to increase the force, force is changed directly. Equivalent mass is no longer recognized as added mass.

If this practice became standard, then some of the discussion in other concepts of this book might need to be converted to the new practice being proposed here. We should normally no longer say or think that acceleration is proportional to the applied force, especially when "relativistic" velocities are involved. In those circumstances, it might be better to say, "Acceleration is 'relativistically' (or simply 'relatively') proportional to the applied force."

The force-factor can still be ignored except when "relativistic" velocities are involved. When v = 10% c, force must increase by only about 0.5%. However, when v = 20% c, then force must increase by about 2%. (Note: this result is the same as with "equivalent mass." However, this new understanding is consistent with the simpler assumptions of **MU**.)

This increase in force F is not caused by a definition or by a formula. The increase in force above the baseline is actually related to the elliptical nature of the afterwarp.

(Note that because the warp is elliptically diminished, then the force above the baseline that is supplied by the afterwarp to cause a particular acceleration must be proportionally increased to achieve the expected acceleration a = f/m, for all m > 0 and f = baseline force.)

Henceforth, $m = m_0$ and $f_e = f \cdot \gamma$, where the FitzGerald force-factor $\gamma = (1 - (v/c)^2)^{-1/2}$. This implies that as the speed of an object with a constant mass of (m) ranges from 0 to c, that its effective force f_e for acceleration (a) ranges from $m \cdot a$ to infinity.

Note that f_e approaches infinity as the afterwarp approaches linear. However, a ray can be considered a (degenerate) ellipse.

"***Relativity Concept***" has now completed the *narrative* that explains (or more appropriately "explains away") equivalent mass.

An Illustration of Natural Relativity

Consider a spacecraft X traveling through *stationary* frame F with parochial speed $\upsilon_0 = 0.2$ c and a continuously emitting dome light that is visible from every direction. $P_x = c = 1$ kcg/s is the parochial speed of the light (with respect to X) of any photon traveling through X.

Q: What is the alien speed $A_X(F)$ with respect to X of a photon when traveling parallel to X through stationary frame F after exiting X?

A: As with all *parochial* speeds, $P_x = c$. As usual, *alien* speed must be calculated. Our strategy will use the fact that the photon and spacecraft X are both traveling in the same direction through F. The question then becomes, "How much faster is the photon traveling than is spacecraft X?"

It is known that a photon in X has constant ***parochial speed*** c. Since we are calculating the ***alien speed*** of the photon with respect to X while it is traveling through F then that would be equal to the parochial speed of the photon, which is c, minus the speed of X labeled n_F that is 0.2 c. Thus, $\mathbf{A_X (F) = c - v_F = c - 0.2c = 0.8\ c}$.

These calculations are not new, surprising, or difficult. However, in forced relativity these considerations are discarded because it is assumed that c is a universal constant and therefore the measuring devices that gave us these consistent values may have lied to us.

These measurements could be a lie, you see, because these same measuring devices are telling us that the speed of light is made constant in every RF by adjusting rather than by supposedly being a universal constant. The fact that we got the same parochial answers could be coincidental and thus irrelevant.

To answer this same question in forced relativity would require the use of the FitzGerald equation to compute the factors of time dilation and distance contraction as well as the synchronicity adjustment since speed equals distance divided by time.

This may seem simple to those who have become accustomed to such calculations; however, it is nowhere nearly as simple as the calculation above. Furthermore, there are only assumptions and no *narrative* for the need to make this calculation more complicated.

Special relativity is a mathematical system where un-natural assumptions must be made to obtain specific values for quantities that cannot be corroborated. The intentional use of "forced relativity" will likely diminish when a simpler system such as this natural relativity becomes available. Natural relativity obtains all of the same parochial results from quantities that can be measured without the un-natural assumptions or uncorroborated results. (The *opinions* above as well as those to follow, will all be justified within about 18 pages below.)

Parochial Speed vs Alien Speed

Warning. What you are about to read may not seem to make sense at first. However, it has been worded very carefully. The meaning will eventually become obvious.

Consider the same spacecraft X traveling through *stationary* frame F with parochial speed $p_0 = v$. Now consider a particular photon p_1 emitted within spacecraft X with an initial *parochial* speed $P_x = c$

traveling in the same direction as X. Due to the speed v of X, this same photon p_1 at the same time but with respect to stationary frame F has an initial <u>*alien*</u> speed ($V_F = c + v$)*.

Upon leaving X, the photon will then be traveling in the <u>stationary</u> frame of F. As such, its new parochial speed will immediately convert* from **alien** speed $V_F = c + v$ with respect to the frame of F [within which it still (but now directly) travels] to establish its **parochial** light speed $P_F = (c + v) – v = c$ in its new stationary frame F.

Notice that the calculation of the speed of the photon p_1 must be adjusted* to its alien speed $P_x = c$ to <u>maintain</u> constant parochial speed $P_F = c$. (The photon is still traveling at the same speed through its new environment of F that it was maintaining through the previous environment of X.) However, it automatically lost the speed of spacecraft X (due to the fixed nature of its now stationary frame) while its parochial speed (through the two different frames remains constant.

* *This is consistent with natural relativity, but contrary to the assumptions of forced relativity that are* ***only*** *valid in that asys.*

This demonstrates that natural relativity and forced relativity are equivalent parochial systems. They both obtain the same parochial answer verifying that the parochial speed of light is constant.

In <u>natural</u> relativity, the ***alien speed*** of light, $V_F = c + v$, where V_F is the **alien speed** of <u>*photon*</u> p_1 (in X) with respect to the <u>*stationary frame*</u> (F), whereas v is the parochial speed of X through F.

In <u>forced</u> relativity, however, the ***alien speed*** is always <u>*assumed*</u> (forced to be) c.

This forced assumption seems to make this version of relativity simpler. However, this faux simplicity comes at a high price because natural relativity has simpler calculations and it also has much simpler Occam.

Correct Physics

Q: Which system represents correct physics?

A: Natural relativity is based on a physical point of view by presenting a physical *narrative* of how parochial light speed is fixed at constant c. Whereas, forced relativity, having no physical *narrative*, is based on a mathematical conversion to the constant frame of all photons by ***assuming*** light speed to be fixed at constant c for alien speed as well as parochial speed.

Therefore, forced relativity is excellent mathematics and ingenious, counterintuitive, assumption-driven physics.

Whereas, natural relativity is grounded *narrative*-driven physics with simple mathematics.

Short answer: both systems are "correct" versions of physics producing the same answers (from different points of view), but natural relativity is ***much*** simpler. This will be demonstrated in multiple ways next.

Remember, due to the newly revealed (in this book) nature of photon travel, any light beam in any version of relativity must always physically be in the parochial environment within which its speed is being measured. This is a trivial identity. The calculations above are very simple when they have been done in natural relativity.

To perform *calculations* in forced relativity (not shown here, but can be found on line), FitzGerald transformations must be used to compensate for time alterations and for manual adjustments (parallel to the eccentricity) of measured distances. Synchronicity adjustments are also common in forced relativity.

Beginning physicists and physicists new to natural physics should calculate parochial speed using natural relativity to verify and strengthen your understanding. Any value for speed v can be used for the calculation such as v = 0.2c.

Constant Frame

The speed of the source does not affect the speed of light. Light instantly transforms to the new speed with respect to the new local warp field or constant frame (CF). Photons must travel at a constant speed of c = 1 kcg/s in order to travel through warp fields and through the constant density space network. Similarly, if the receiver is moving, the photons must transform their speed to that of the local warp field of the receiver instantly upon arriving there.

This narrative explains how light always travels at speed c = 1 kcg/s with respect to its CF. In fact, this is the reason the parochial (local) warp field can be called the constant frame of light.

If these calculations of alien c seem superfluous, this is simply because forced relativity is unable to justify or explain how alien c can be assumed constant, thus it never tries. "This is just natural behavior."

(This oversight is forgiven, of course, since many people were misled by a broad interpretation of the Michelson-Morley results.) It now seems (to them) unnecessary even to consider this point.

"When particular assumptions lead to conclusions that do not contradict those assumptions, then the assumptions *must* be valid, you see."

I know … right?

The problem is that particular assumptions lead to particular asies. This tweak is a simple perversion of principium axioma by affirming the consequent.

In the case where c was assumed constant, the conclusions are only valid in an asys where c is constant. Since, by *narrative*, alien c is a variable, then forced relativity is only ***physically*** valid in a parochial setting; it is only ***mathematically*** equivalent in alien settings (Conversion formulas must be used to obtain solutions that are physically valid.)

Constant kcg/s of Photons

There are 2 underlying principles that keep c constant in its (parochial) CF. Of course, the expansion of the geometry of space is significant. However, the determining factor for parochial light speed in a vacuum seems to be the number of kcg/s that a photon is able to realize.

If all photons are alike and all motron spans are structurally similar then it would be surprising if the average kcg/s were not constant. This seems true because there are so few, if any, factors that may be able to affect it.

The average length of motron spans, called "span average" could also be relevant. Denser fields would presumably have smaller span average and seem to slow passage of radiation. It is known that light travels slower through air than through a vacuum and even slower through glass than through air. However, in a vacuum, the varying of span average caused by varying frequency of the radiation is negligible.

Note that c is never measured in spans per second because of span variability. (It does not help that it is also not visible.) However, c could be related to the average parochial span per second. This average will be called the "spoke."

We know "spoke" is held constant by the motion process; since every span value is so tiny, then the number of spans per second would always be extremely large. Because of this great number of spans per second, then making spoke an average (per second) of a great number of tiny spans would increase the accuracy considerably.

It is also expected that under most circumstances those photons with higher kcg/s travel faster than those with slower kcg/s do since higher kcg/s implies more spokes which results in greater speed.

One consequence of this is the dispersion of light traveling through a prism. However, this variability is also probably negligible in a vacuum.

Because c (in a vacuum) is so constant and so accurate in all circumstances where it is measurable, and because of the relativity conjecture, it has become the standard. If the measured parochial speed of light in a vacuum were ever significantly different from the assumed value, then some other factor would always be assumed complicit and not the consistency of c.

(Since it is unnecessary to adjust calculations to maintain a constant c, then the Doppler Effect for radiation works exactly the same way as the Doppler Effect for sound in any situation where the speed of sound is considered constant.)

Doppler Shift Observers

Consider, for example a light wave W traveling toward stationary observer O_A. With respect to O_A, W will be traveling with a *parochial* speed of progress, speed c. Because of this, the frequency, f and wavelength λ will be ***normal***.

Now, consider a second observer O_B to be traveling in the same direction as W with speed v_B. With respect to O_B, W will be traveling with a decreased* alien *speed of closing*, $c^- = c - v_B$ ***toward*** O_B. Because of this*, the frequency f will be decreased and the wavelength λ will be ***red*** shifted.

Finally, consider a third observer O_C to be traveling in the opposite direction as W with speed v_C. With respect to O_C, W will be traveling with an ***increased* alien speed of closing***, $c^+ = c + v_C$ ***toward*** O_C. Because of this*, the frequency of W will be increased and the wavelength λ will be ***blue*** shifted.

* *This is consistent with natural relativity, but contrary to the assumptions of forced relativity that are **only** valid in that asys.*

Summarizing these two effects (sound and radiation), we might say that the change in frequency is always caused by a change in progress of the vibrations. It does not otherwise matter if the source speed changed, the receiver speed changed, or if the distance changed due to expansion.

Since there are two different narratives (sound and radiation) for the measurable results of the Doppler Effect, then it is impossible to use this effect to prove whether the density of the universe is independent of motion. Hence, it cannot be used in any attempt to prove that alien c is also constant. As with all "proofs" of constant c, the consistence of c must ***first*** be assumed. The <u>assumption</u> that c is a universal constant can then trivially be used to "prove" that c is a universal constant.

Constant c Analogy

If the following analogy does not make sense for you, just speed it up until it does. This may not happen, however, until you reach speed c. Use a jet skate-board or jet roller-skates (depending on your age) if you have to.

Suppose in trying to get to an appointment on time you are walking elbow-to-elbow with a large group of people through an airport at uniform "crowd-walking speed." To save time, if your entire group could maintain their crowd-walking speed while walking onto a conveyor belt "people mover," your uniform speed would seem to increase as seen by people looking down from the second floor overlook.

These same people looking down might also notice that the distance between people has increased. The mechanical pedometer / speedometer on your hip would register no change in speed since your feet and legs are still moving as before. There are two ways to "view" this phenomenon.

1. While your parochial speed remained constant, your "air" speed increased as you added the conveyor speed. This got you to your appointment early.

The discreteness of the transition automatically increased the distance between people as they stepped onto the conveyor.

2. Your speed remained the same and anyone who is not waking with you is unable to judge your speed because they are not stationary with respect to your group so that their sense of distance and their sense of motion and time are no longer trustworthy.

The distance between people has not changed. This misconception is caused by the presumed length contraction. There was similarly no time saved by the people mover. This misconception is caused by the presumed time dilation.

However, you did arrive early for your appointment, due to your presumed adjusted synchronization.

The first view is a model of natural relativity while the latter is a model of forced relativity.

Both of these views are reasonable and both will yield the exact same correct calculations once transformation factors are calculated and used. However, the first view is much simpler, much easier to understand, much easier to communicate, and ***transformation factors are unnecessary since no transformation occurred*** from this first point of view (asys).

How Forced Relativity Changes Measurements

For the purposes of this discussion, we will suppose there is a space ship called SSh capable of relativistic velocities. SSh, with speed v passes space station SST that is presumed stationary.

According to forced relativity, this speed of SSh shortens the length dimension on SSh in the direction of motion and dilates the time dimension on SSh as seen by observer 2 on SSt. These changes are not noticeable to observer 1 on SSh. The proper frame of SSh has not changed.

Calculations show that, in order for c to be assumed constant, the observer who is assumed stationary must see the measurements that are being made on the moving ship (if it were possible to see then at all) as altered by some mysterious but necessary distortions of the environment. (This cannot be verified physically since SSh is moving too fast.)

(We say, "must" because how else do you explain how the "speed" of light could be assumed unchanged when "progress" has changed measurably until this, too, seems explained away by supposed loss of synchronicity.

These measurements are impossible to make, of course, so we must assume that if we could make them then we must get what we have assumed!

Since the space network is not detectable (yet), it is not possible to tell which of these two observers is actually moving. Thus, it is still possible, using the principle of classical relativity, to reverse their roles.

When we do this, we get "mirrored results" (the same results in reverse between SSh and SSt).

<u>Explaining Time Dilation</u>
<u>In Forced Relativity</u>

The best way to understand how time is thought to dilate on SSh as seen from SSt is to remember that events occurring on SSh that seem normal to passengers on SSh, are seen by them as images traveling at speed c still within SSh. In particular, a photon traveling any distance in (CF) SSh would have speed c regardless of its direction of travel.

There are several ways that an observer on SSt could interpret the (lateral) motion of the photon on SSh as SSh speeds past SSt. If the speed were calculated as v + c, it would exceed the maximum speed, c. However, before that can happen, these images from SSh must transform. This is because light travels with speed c with respect to the medium, or field, in which it is traveling (its CF).

The slowed images would look like a slow-motion video as seen on SSt. Since all images are slowed, it would appear that even the clocks on SSh were running slowly.

This idea is unrelated to time or any possible dilation thereof. In order for the image to be seen on SSt, the images would actually have to travel to SSt. As the images enter SSt, they must slow to maintain a (new) constant speed c with respect to their new CF.

Time does not appear to dilate from the perspective of SSt unless we <u>force it to</u> (when making alien calculations) by ***supposing that the newly assumed speed of light must be the same as the previous calculated speed.***

This explains forced time dilation. Since time in this case cannot be measured, it must be calculated. The ***only*** discrepancy between forced relativity and natural relativity occurs because of alien measures, which must always be calculations but are called measures. Otherwise, we obtain the same result both ways.

To summarize, the change is forced if we assume that the alien speed of light is still the same as before even if we now either …

(a) Calculate it with respect to a non-relevant reference frame (such as occurred in the Michelson Morley experiment) in which the photons are not traveling, or

(b) Calculate it using time dilation or length contraction in order to obtain the *assumed* value of c that is no longer relevant.

We get the same value as before because we assume the measuring instruments have changed so we use a formula that calculates the value based on the *assumed* distortions in the non-relevant measuring instruments.

The change is *natural* if ***we find that the speed of light is still the same as before*** when we now calculate or measure with respect to the new reference frame because the speed of light has adjusted itself to the new frame in which it is now traveling. All measuring devices still work (unchanged) and there are no synchronicity problems.

Who knew relativity could be so simple. Of course, it will seem more complicated if we must explain our reasoning or if we must now perform the calculation both ways (as we have just done) as a check of either accuracy or relevance.

When is forced relativity still relevant?

To illustrate the importance of forced relativity, even when natural relativity is available, consider the case where SSt is stationary. We can now convert forced relativity argument above to say, "Absolutely," that it is the motion of SSh that makes its clocks appear to run slower as witnessed in real time on SSt. This sometimes makes it easier to make the alien calculations between two "third party" locations if we do not care what is "really" causing the disparity.

Mathematics often seeks alternative methods to circumvent missing data or difficult physical circumstances.

The Michelson-Morley Experiment

The Michelson-Morley experiment of 1881 and others like it, attempted to determine the motion of earth by measuring the differences in the speed of light in different directions. The failed results of the Michelson-Morley experiment are purported to have proven (1) that the speed of light is "always" constant.

While this conclusion (1) is certainly one possible conclusion, it is not the only (or the simplest) possible conclusion. A careful analysis of all such experiments shows that they only establish conclusion (2) that the parochial speed of light is constant.

If conclusion (1) that the speed of light is always constant had been proven, then conclusion (2) that the parochial speed of light is constant would also have been proven. However, conclusion (2) does not imply conclusion (1). Having established parochial constancy, then these negative experimental results would be expected since the parochial frame never changed throughout the experiment.

The entire experiment was performed in the reference frame (environment) of earth. The light beams, though within the solar system, were traveling in the parochial environment of the earth, not the solar system so light speed did not respond to changing orientation of the experiment with respect to the earth.

Most of the final conclusions of motion experiments are the same in natural relativity as they are in forced relativity. It is mostly the simplicity and understandability of the explanation that has changed.

Alien measures must still be calculated, but since alien measure (calculation 1) has changed to alien measure (calculation 2) with the same parochial value instead of being assumed constant, we still get the same results, but with no distortion of instruments.

In mathematics, the choice of coordinate system can sometimes have a big influence on the simplicity of the computation. However, when done properly, it should not alter any "physical" conclusion unless we have a specific reason to enter the mathematical realm of "equivalency."

"Proof" of Special Relativity

"General relativity" has never been completely tested. Natural relativity reduces much of this conjecture to "irrelativity."

However, "special relativity" has been thoroughly tested and it has passed each and every test. Does this not prove its correctness?

Unfortunately, no, it only proves its consistency. This rejoinder will be proven now.

The primary difference between "special relativity" and "***Relativity Concept***" is the assumption in forced relatively that alien c as calculated would have been the same value with respect to our parochial measuring environment if it were possible to measure it with our alien devices.

This sounds a bit presumptive, does it not? The parochial measure of alien c is naturally "c" with respect to its parochial frame. The corresponding value can only be assumed in the alien frame because moving instruments are incompatible with stationary observers and therefore alien "measures" are impossible for us to actually measure or even see.

Alien measures must always be calculated since they are impossible to measure directly. These calculations can be made assuming a universal constant c. This mathematical reasoning, however, cannot then be used to prove (circularly) what was already assumed … that alien c is "physically" a universal constant.

Like "***Relativity Concept***," the "***Special Theory of Relativity***" is inductive and cannot be proven deductively. Any weaknesses or inconsistencies within forced relativity have apparently been purged.

The same will optimistically be soon accomplished within this new (natural) relativity. It will then require Occam's law of parsimony to ferret out the stronger theory.

Q: What about experiments that "prove" time dilates on rapidly moving space satellites?

A: Notice that the idea that time dilates aboard a moving space craft is based upon calculations that are considered measurements. ***All*** such "calculations" performed in the past have assumed that light speed is a universal constant. This assumption continues to the present day.

Therefore, yes … if you assume that alien c is a constant for any alien observation during any interval of time, then calculations based on that assumption will show that time is dilated and distance is contracted during that same interval.

You will get the opposite results during any interval when you assume the opposite. This is a simple application of principium axioma.

Gray Reasoning: When you assume that light speed did not slow down upon leaving the moving frame, the ***bizarre*** results of your calculations are telling you, "Please think again." "The only basis for such a "*gray*" assumption would be if "*gray*" time somehow sped up and particularly "*gray*" distances inexplicably shrank." "Are you really willing to go to this "*gray*" place?" This reflects relativity of course because "*grayness*" is also relative.

Reality Check

Q: Suppose we wanted to track "aging" on a rapidly moving satellite. Would it be possible to create on earth a relativity clock that ran as slowly as a "time-dilated" satellite clock?

A: Let's find out. First, use the speed of the chosen satellite to calculate the time dilation that would result from its motion. Then adjust (slow) the "ticking" rate (percentage) of the relativity clock accordingly. However, it would also be continually necessary to add seconds or fractions of seconds to the (stationary) relativity clock in order to maintain "manual synchronicity with the (moving) satellite clock. Having accomplished these adjustments, we would discover three interesting facts.

Fact 1:

A higher percentage was added to the clock for the "synchronicity adjustment" than was subtracted for "time dilation." This difference is purely mathematical, however. If time dilation reduced time passage on the original clock, then it would requires a greater ***percentage*** boost for synchronicity to increase the <u>lesser</u> time by the same amount that the (original) <u>greater</u> time would have been decreased by time dilation.

For example, a 10% decrease of 100 produces 90, whereas a 10% increase of 90 brings us back up to only 99. Those who struggle with percentages might consider performing calculations in minutes and seconds when possible (somewhat similar to what was done here earlier) instead of using percentages.

Fact 2:

The "time dilated" clock would always be running slower than ordinary earth clocks, but after each synchronicity adjustment it would show the same clock time as ordinary earth clocks.

Fact 3:

This cancellation of the time dilation / synchronicity effect helps us to see that it is not necessary to adjust for time dilation because c automatically adjusts to remain parochially constant. Therefore, <u>time does not dilate so there is no need to adjust it (or synchronicity) from</u>

this point of view. This also tells us that from this point of view these two effects are the same effect in reverse.

Objections to the Reality Check

Some might try to claim that the reality check is incorrect because the relativity clock is not actually moving, therefore the synchronicity loss would be zero. This would mean that the moving clock really is running slowly.

Nice try, but if you model a phenomenon, then you must be consistent. If you remove the synchronicity correction (because the clock is not moving), then you must also remove the time dilation. The only reason the model was running slowly was because we assumed it was running slowly due to its presumed motion. With that presumption, then the clock must also be adjusted for its presumed loss of synchronicity based on the same presumed motion. The conflict is caused by mathematics, not motion.

Critical Comparison

Suppose photon X travels 20% farther in the same time and same direction within a moving system, compared to an adjacent photon Y that is traveling within a stationary system. Instead of assuming universal c for all observers, it is instead possible to assume universal time t and equivalent measuring devices for all observers in all situations.

(Please note that, for most people, these assumptions of universal t and d are far more universally acceptable, demonstrable, and understandable than the assumption of universal c.)

These assumptions will allow us to calculate any difference between parochial c and alien c. Furthermore, all results can be well documented.

It might (but should not) surprise us that measurements obtained from careful calculations now confirm that:

- Alien t and parochial t are exactly (and demonstrably) equal when measuring the passage of time for any event within these two systems. (Reasoning to follow below.)

- In addition, since *parochial* c is exactly the same in both systems (as it is in all systems and is not in question), then measuring devices do not shrink within rapidly moving reference frames under these conditions. (Equal velocities and equal times imply equal distances.)

- Therefore, the difference between parochial c and alien c can be confirmed by actual measurement.

- We see that photon X went 20% farther in the same amount of time within its moving (alien) environment than photon Y did in its parochial environment. This tells us that the environment of photon X was traveling in the same direction as X and at 20% of c. Therefore, the alien speed of X was 120% of the speed of Y.

- This proves that alien c is affected, but time, length and synchronicity are not affected by moving reference frames.

- This proves conclusively (and finally) that c need NOT always be a universal constant. Only if a decision is made to adjust c to keep it (force it to be) constant.

It might be difficult to find funding for such an obvious experiment, however. Please read this proof again if you did not "get it" the first time. No insinuation intended.

Fans of special relativity should answer these same questions using that asys. Note that you get the same <u>parochial</u> result as well as any other results that can be physically verified. Also, note that the calculations are unnecessarily complicated when assuming constant c.

There are no surprises and <u>no tricks</u> in the preceding proof. If you assume universal t, then you have assumed that alien t and parochial t are equal. Using that assumption to calculate alien t results in equality between alien t and parochial t. You <u>always</u> get back what you assume.

This situation where alien t and parochial t <u>become</u> equal should look familiar. This asys works <u>exactly</u> like the familiar asys of forced relativity where c was assumed a universal constant and alien c and parochial c <u>became</u> equal.

This comparison makes it obvious that the unusual results of forced relativity are not caused by motion, but by changing the asys. Therefore, by principium axioma, the strange results of special relativity are only applicable in the mathematical world of "constant c." Trying to understand the universe within that asys is even more difficult than trying to understand the solar system within the geocentric asys with its cycles and epicycles.

The universal c system is mathematically correct, however, so it <u>can</u> be used for calculations concerning "constant-c-world" but <u>not</u> for insight into other asys (such as our "normal constant t" universe). Just because it is done all the time does not make it easy or necessary.

Natural Relativity Proof

For reinforcement of these concepts, let's run the Forced Relativity Proof again, but this time assuming a universally constant distance scale, a universally constant time scale, universal synchronicity, and universally constant ***parochial*** speed of light, c.

We would need to calculate the actual speed of the spacecraft toward us. Then subtract this value from c. Using this as the alien value of c will allow us to calculate all distances and times in the moving frame accurately.

This can be verified by the still correct instruments aboard the spacecraft. There is now no evidence of length contraction anywhere and no time dilation either. The distortions that were thought to exist were actually ***caused*** by relativity *assumptions* and ***not*** by *motions* of the spacecraft!

How Do We Deal With Delayed Images?

The primary difference between general relativity and special relativity is that in "The General Theory of Relativity," systems are studied which are allowed to accelerate. The only external way to observe an accelerating system is by use of some sort of carrier wave. This might seem like a slow-motion video, but it would be real life.

To deal with this, we must decide how to think of delayed images. This is very similar to using look-back time. With look-back time, we always see exactly what happened, but sometimes much later than it actually happened. Should we think of it as something that is happening now as we observe it happening, or as something that happened "back then?"

There are advantages and disadvantages to both methods of thinking. We know when the event "seemed" to occur, but can usually only put an approximate date to when the event "actually" happened. We do not know what has been (local time) happening there since the event, or even what is (local time) happening there "now." We will use the expression "***local time***" to refer to the time of an event as measured at the site of the event.

Overall, it is usually simpler to think of the viewed events as current. For example, the supernova explosion (that left, in the night sky, telescopically visible remnants we now call the Crab Nebula) is said to have occurred in 1054 A.D. That is when the brightness of this event was first seen from Earth as a supernova visible (appearing as a very bright star) by everyone (with adequate vision) around the world (at the appropriate times when the object was above the horizon for them) *even in broad daylight*.

A very similar analysis can be applied to relativistic events. It is usually simplest to think of the delayed images being witnessed (or imagined) as if they were current. In this way, we get the same understanding as with look-back time.

The Now-Universe

At any point in time, there is only one universe. **MU** will refer to it as the "now-universe." Objects and events that exist or occur ***one light year*** away from us can only be seen or witnessed by us (as they were at that time) exactly ***one year*** later in earth time. A telescope is only necessary due to the great distance of one light year. Telescopes makes objects larger and thus "seem" closer, not sooner.

The time of the event occurrence was called, "now" in local time. The event can be seen from earth at "now+1 year" in earth time. It is likely that this notation will be of value primarily for predictable events.

Objects or events that exist or occur ***one light second*** away can only be seen or witnessed by us exactly ***one second*** later at our "now."

This is another reason we have no need for a time dimension. Local events in history may be separated by time. Current events in the universe may be separated by distance. Events in space-time may be separated by both distance and by time, but those are two different separations. These separations divide space into four quadrants that are labeled according to our time and place.

1. Earlier, Closer
2. Earlier, Further
3. Later, Closer
4. Later, Further

We can only exist "here" on the boundary between "closer" and "further." We can only exist "now" on the boundary between "earlier" and "later."

We are able to observe "then" and "there" at different distances from "here", but only when "then" is in the region between "earlier" and "now." The exact occurrence of "then" will shift "earlier" as "there" moves "further" away.

We can affect no changes on the past being viewed, process no sensory evidence beyond the visual, and bring no artifacts back to the present other than visual records. To be seen, every event of the past must be viewed at a specific time (regardless of where we are) and none can ever be viewed "live" a second time. The ***effects*** and even the ***occurrence*** of a protracted event can sometimes reverberate through the universe for an extended span of time, however.

One other universe of interest is the "look-back universe" or preferably the "visible universe." The stars we can see in a telescope can be separated by hundreds of light years and galaxies by millions. This is not only a great distance; it is also a great separation in time. The light we see simultaneously from adjacent stars may have been emitted at times separated by hundreds of years. These look-back images are all part of the same "visible universe."

The visible universe of a particular date and time only exists or existed at that specific time. After that, we can only talk about it or look at pictures and silent videos. Sound does not travel nearly as fast as light and not at all through a vacuum.

The distance variable has 3 independent coordinates with which the contents of "here" can be changed by our motion.

The time variable has only 1 independent coordinate for which the value is always "now."

There are many things that can separate people in real life such as cultures, attitudes and mores. We could even establish a 5-D space-time-temperature universe.

However, should we? Why should we? When might we find that useful? Might a graph prove more useful? Might two or three dimensions at a time prove more manageable?

MU will only use a time dimension when it cannot easily be avoided. Unless otherwise specified, natural relativity will always assume the "D3-now" universe. "Now" can be any preferred point in time. A specific "now universe" refers to the entire visible universe of that time.

Events that are more distant occurred longer ago in the look-back time of the past. Meanwhile, the aging events of lookback time advance toward the visible universe as the visible universe irresistibly and quite uniformly progresses into history.

A specific D3-now universe might be D3-(7/04/1776). At the time of this writing, most of the "contemporary" objects and events of that now-universe (including many of our ancestors) are currently available (live and in color but from different perspectives) about 240 light-years away in any direction. I will leave it to the curiosity of each reader to figure out what else, where and when it might be present, (if not quite visible) in any specific D3-now universe besides our ancestors. This could become a fascinating pastime for a few very special people. They might be called, "Now-seekers."

How Forced Relativity Shortens (Contracts) Length

Let us return to the discussion of SSh and SSt from "How Relativity Works." We have learned from the discussion of <u>forced</u> relativity that the speed c is (assumed) the same in every frame, including SSh and SSt. In addition, clocks were seen to be running slower on SSh as seen from (stationary) SSt. These facts can be used to compute lengths of objects parallel to the line of motion. The argument goes as follows.

Suppose, while traveling at a high rate of speed, we measure a beam of light traveling, within our parochial system, past us in our direction of travel and discover that (as it always does) the light has traveled the exact distance that it should in the given amount of time. If we assume that, the alien speed of light is constant so that due to our high rate of motion, our clocks "must be" running slowly, (Please see how time dilates above.) then the light must have traveled longer (because more time supposedly passed) than what we measured. Thus, the light must have gone further than we measured. Therefore, we must conclude that our measuring devices must now be shorter than they seem to be when moving at slower speeds. (How is this not like the Emperor's New Clothes?)

This "proves" that these lengths (under those conditions) were shorter on SSh than they would be on SSt. By reversing the argument, (again using the classical "principle of relativity") observers on SSh would draw the same conclusions about events and objects at SSt. These length contractions and time dilations conform to the requirements of forced relativity.

Alternative Realities?

There is one big question that we must all face honestly. Note, however, that there are two correct answers, so in this case "correctness" (unlike truth) is a ***decision***.

There are only two possibilities, either c is a universal constant or else it is not. Deciding A) that it is a universal constant results in special relativity. Deciding B), that c is only a parochial constant results in "natural relativity."

Q: The big question is, "How does this decision change reality?"

A: The big answer to this question determines which asys we are choosing.

a) We can (and have) proven that special relativity is a complex mathematical system that is artificially related to the real world in an abstract way. When properly translated, this system does allow computations using "conversion" factors.

b) "Natural relativity" allows us to see the world in its simple, "natural setting." "Conversion" factors are unnecessary.

Relevance of "Vamos"

Q: Is length contraction, from special relativity, related to shortening of vamos in **MU**?

A: No, contraction is across the board parallel to the direction of travel, whereas some vamos is longer and some is shorter in every direction. In addition, vamos is a physical reality (with a physical *narrative*) whereas length contraction (with no *narrative* possible) does not physically occur in any alien system. Length contraction is only a mathematical conversion (calculation) between two inconsistent systems (where, due to the assumptions being made, conversion is required to compute alien measurements between systems).*

* *Many physicists disagree with this assessment. Some would argue that any perception is reality that must be understood within the boundaries imposed by the assumptions made.*

This disagreement, however, confirms the point being made in "natural relativity" that all conclusions within any asys apply only to that asys. Therefore, c is universally constant only in a mathematical model where it has been assumed a universal constant, not necessarily in other models of the physical world.

How Mass Increases In Forced Relativity

It is even easier to physically understand how and when mass must increase with speed. At higher velocities, SSh (from the previous discussion of forced relativity) requires greater force to achieve the same acceleration, a.

If ($f_1 = m_1 \cdot a$) and ($f_2 = m_2 \cdot a$), then $f_2 > f_1 \Rightarrow m_2 \cdot a > m_1 \cdot a \Rightarrow m_2 > m_1$. Thus, the proportionality "constant" called mass in the force equation has increased because it now requires more force to give the same object the same amount of acceleration.

Q: If we now assume that SSh is stationary and SSt is the one moving would we then discover that $m_1 > m_2$?

A: No, we are not talking speed here. The "increase" in mass is caused by acceleration. Achieving acceleration requires applied force. The object that is accelerated is the one that "increases in mass." This is the main reason it is assumed that force (or the resulting acceleration) increases mass. Please see discussion of equivalent mass above. Make sure you understand why equivalent mass can no longer be tolerated.

Proportionality between Matter and Mass

The above discussions imply that the amount of mass need not always be proportional to the amount of matter unless we count the matter corresponding to the mass that is "equivalent" to the force that supposedly caused the acceleration.

This is ironic. We often distinguish mass and weight by noting that, since weight is dependent on location (such as on earth or the moon or even the microgravity of space), it is not always proportional to the amount of matter.

Now we see that mass has a similar problem with motion. To compensate for this (mathematical) increase, the added mass is called equivalent mass.

In natural relativity, increased force is necessary to compensate for the elliptical afterwarp. Therefore, it is unnecessary to add equivalent mass to make the effective force equal to the (rest) mass times the observed acceleration. It is unnecessary because they are already equal.

Thus, in natural relativity, effective force is always proportional to the acceleration.

Coordinates of Angle-warp

The lines of force that cause overwarp converge in the direction of acceleration. Thus, an object can be 1) accelerated into orbit around a planet and simultaneously be 2) accelerated under the attraction of gravity.

Such an object not only can experience Lorentz–FitzGerald* contraction (named after physicists Hendrik Lorentz and George Francis FitzGerald) parallel to the direction of travel, but also perpendicular to the direction of travel. It would seem, then, that these two perpendicular forces are simply the coordinates of the angle-warp.

Note that the direction of the speed vector in one of the coordinates can be constant while the magnitude changes. The direction of the other speed vector can be changing even if the magnitude remains constant.

* *Even in natural relativity, some of the language of forced relativity can still be used for the benefit of those of us who have learned these concepts within that environment.*

Narrative for Parochial Constancy

It is time to explain how the speed of light is able to adjust its speed as it travels from one system to another so that its ***parochial speed*** always remains constant c.

The truth that many people are missing (including some physicists) is that it does **not** change.

If a photon beam with speed c travels from a stationery frame, with frame speed $v_0 = 0$, to a moving frame, with frame speed v_f, in the same direction as the photon, it is ***not necessary*** for the "new" speed of the photon to physically ***slow*** from a new speed of $c + v_f$ to the speed of c.

It is sometimes explained as a slow down because this **is** what happens *mathematically* and this is how many physicists think of it. We should certainly question that notion.

If the speed really ***did*** change then we could rightly wonder how that might happen. Furthermore, how would you ever explain the negative result of the Michelson-Morley Experiment?

It is now necessary "narratively" to explain what happens physically so we can understand why c is a parochial constant but not an alien constant.

When a photon travels from the first frame to the second frame, the photon continues with the exact same speed. We have already seen by narrative that the ***parochial*** speed of light through any system is always constant c.

The rate of time passage is always the same in all environments. We are calling this, "constant t." Since all distance measurements are unchanged by motion, "constant d," then **no** measuring devices have changed.

To locate objects in the new environment and remain current in our viewpoint, it is generally desirable to change coordinates to the new system (or frame) immediately as it changes. It should be stressed that discussions like this generally use the word "new" to refer to the current frame. The word "old" generally refers to the previous frame.

If, however, we continued to consider the speed of the photon with respect to the old frame (in which it is no longer traveling), then it would seem, that the photon is now traveling faster (at $c + v_f$) than its speed of c just before the transition. If we then switched our reference frame to the new system, then that would seem like a "slowing" from $c + v_f$ back down to c.

Thus, nothing in this "new" frame has really changed, and nothing needs to change to explain why the speed of light changed, because it did ***not*** change. This should make it easy to remember. "Every photon tends to maintain its parochial speed when transitioning into a new environment." Due to its seeming simplicity, it is not surprising that, in forced relativity, the unmeasurable ***alien speed*** is also *assumed* (forced) to be c. However, there is no narrative for this assumption.

Narrative for
Motion of a Massive Object

We will next consider the motion of ***massive*** objects through the space network. Massive objects have their own momentum. They do not maintain constant speed through the space network. Therefore, when a massive object is overtaken by a rapidly moving frame, it must either physically increase its velocity or travel at a slower speed through its new frame. This is the reason passengers in a moving vehicle can "feel" any significant change in the velocity (direction or speed) of the vehicle.

Momentum with respect to the old frame is automatically maintained with respect to the ***old*** frame. Thus, since the mass m is fixed, then, the velocity with respect to the old frame must also be fixed, and since the parochial speed is not maintained automatically (as with photons), then the parochial speed v_0 must be maintained physically to prevent its alien speed from diminishing.

With massive objects the alien speed is sometimes more important than the parochial speed. The net (alien) speed of a massive object with ***fixed*** parochial speed v_0 through a frame moving with parochial speed v_1 will generally equal $v_0 + v_1$. However, the alien speed of the frame with respect to the object is $v_1 - v_0$.

If the parochial speed v_0 is not fixed, however, v_0 could diminish to the alien speed v_1 of the new frame. The <u>loss of speed</u> would then equal the alien speed of the original frame with respect to the new frame $(v_0 + v_1) - v_1 = v_0$.

Before we proceed, we will emphasize that alien changes do not (and cannot) normally cause parochial changes. This change of alien speed without parochial change has two ramifications.

- Technicians observing this transition are able to say that the parochial speed of any massive object must be constant because it's <u>measured</u> speed before and after any transition is always the same.

- There is always a Doppler shift caused by the stretching or shrinking of the transition progress between the cogs from the first to the second reference frame.

The narrative for constant parochial c has already been given. The ***parochial speed*** of every massive object is still constant as it travels from one frame to the next. However, for massive objects the parochial speed is always <u>*less*</u> than c.

We have now established the narrative for why alien c must change from one frame to the next. It is so simple that it is easy to understand why it seemed so complicated due to its significance.

Does it not seem foolhardy to adjust measuring devices (or their results) that <u>*are*</u> still functioning exactly as before and still giving correct readings? Is it not possible that the only thing incorrect was our expectation of changed speed from one ***parochial*** frame to the next?

However, massive objects traveling to a moving frame (with speed vector v_f for the frame) always maintain their speed and direction with respect to the parochial frame. Circumstances generally require that speed and distance traveled be re-expressed with respect to the original frame. ***Therefore, the original speed vectors have all changed to*** … [$v_{new} = v_{old} + v_f$] … ***within the new frame but with respect to the original frame.***

"*M*assive *m*atter *m*aintains its fra*m*e." This sentence provides the "***m*-rule**." The m-rule implies that when the frame is also moving that the resultant of the transfer will include the vector speed of the frame plus the vector speed of the object within the frame.

Light speed is no different except for "pc." For us as physicists, it may take us some time to help everyone understand that "c" is a parochial constant; it just masquerades as a universal constant." For those who are *not* always "politically correct," "pc" could stand for "parochial constant."

We have just experienced the merging of two systems where very little changed in either. The speed of a photon ***seems*** to change when it enters a new frame only if we continue to consider its speed relative to the old frame.

However, because photons are monopons, they continue to travel at exactly the same speed (through the fabrons of the new frame) as they ***were*** traveling (through the fabrons of the old frame), exactly 1kcg/s in both cases. It is only the frame velocities that are different.

To help us remember how monopon travel works so that we will never make computational errors of this type, just remember the following. Photons only travel via the local frame.

The Speed of Light

Though this may seem colloquial,
"C" is always parochial.

Therefore, photons only travel through the local frame. Thus, since c is always parochial, then c must always be measured with respect to its (local) reference frame. Hence, the measured value of c will always be the parochial value of exactly 1 kcg/s. The ***alien*** value, however, must be calculated.

Might it be possible that, *not knowing how momentum p was maintained*, we overreacted to the known consistency of parochial c by assuming a similar consistency in unmeasurable alien c? "We were just trying to keep it simple."

This completes the *narrative* for the constancy of parochial c.

Imports of "*Relativity Concept*"

Changing speed is called acceleration. Thus, warp fields are capable of changes in both acceleration and in speed (often simultaneously). This is a good indicator that afterwarp is just as real as the momentum it maintains.

We can now say that all motion is relative and all motion is also absolute. However, it would be more forthright (less contrived) to say it this way. All motion is absolute to the space network; it can be considered absolute to its parochial frame; it can be considered relative to any frame. Every subnetwork can be defined by an appropriate coordinate system or boundary container.

To avoid further confusion we must clean up our language. The intention was to find a positive way to say, "Motion can only be considered absolute in a relative way." The disallowing of "absolute" motion was caused by the decision to disallow a "fixed" space network when the presence of aether was contraindicated by an overly-broad interpretation of the Michelson-Morley Experiment. It is interesting that this one misdirection has led to so many unsolved problems in today's physics. Fortunately, this book provides solutions for most of those problems. It is also very likely that many others are closing in on this same conclusion even now.

Whenever we now say, "All motion is relative," we must understand the hidden assumption. Saying it is relative does not necessarily mean "relative to each other," or "relative to us." It just means that all motion must be understood relative to some reference point (that must be considered stationary).

To translate into our current setting, we will take this to mean that all motion is "naturally" relative to its parochial frame in which it travels, in which it can be measured, and in which c is constant.

We can still make alien calculations between reference frames. There are two strategies that work.

1. We can assume, mathematically that c is a universal constant and therefore distances, time, and synchronicity must be adjusted when making alien measures between alien system, or

2. We can assume by *narrative*, that alien c must have a new value with respect to the old system since its parochial value always remains constant c.

Author's Note

Physicists were investigating systems in which the alien speed of light was changing from one system to the next. Yet a "double blind" (Michelson Morley) test seemed to suggest that ***the speed of light might be a universal constant*** that they dubbed "c." It was found that by mathematically altering event synchronicity along with the measurements of the two components of speed (distance and time) a multiplex system could be developed that would allow computations in this strange world of constant c.

Since the main thrush of physics is the making of predictions, then "constant c" world was mathematically developed and the guidebook was called "The Special Theory of Relativity." Now that we know how light travels through the fabric of space, we see that its motion is always parochial.

Strangely, experimentation has proven that at least the parochial speed of light is always constant. This was considered unlikely because the expansion of the universe should cause any parochial motion to increase due to the stretching of space. (The greater the stretch, the faster the travel.)

We now see that the Michelson Morley test was biased, by its constant parochial frame (earth), toward universal c. Constant <u>parochial</u> light speed now implies that the <u>alien</u> speed of light, which is impossible to measure directly, must change from one moving system to the next. This change depends only on the motion of the new frame.

This, at least, seems likely; we face three options for making computations involving motion.

1. We could ignore all modern discoveries and continue using "forced (special) relativity."

2. We could assume that the blind test analyses (claiming c to be a universal constant) were interpreted incorrectly and adopt a far simpler system called "natural relativity" for dealing with high-speed motion.

3. We could assume that the blind test analyses were correct but develop a mathematical system to simplify calculations called "artificial-absolute motion." This would artificially force the simplicity-advantages of natural relativity. This artificial system would then be indistinguishable from natural relativity except in its philosophy.

Options #2 and #3 together suggest that regardless of which theory is physically correct, the procedures of natural relativity can be used to simplify calculations significantly whenever they may involve relativistic velocities. We no longer need ever be concerned whether time may have dilated, distances may have shrunk, or synchronicity may have been compromised.

We need no longer consider the speed of light to be a universal constant. Henceforth, constant c should refer to parochial light speed that we have long known to be constant.

HoHWGH

I would like to take this "time-out" to express my personal opinion concerning the "History of How We Got Here." This can be abbreviated "HoHWGH" for short, pronounced "Hoe-Huff" or "Hoh-Wuff" if you prefer.

To this point in **MU**, we have uncovered many seeming oversights in the development of physics. By laying these concerns out, however we soon discover a common thread.

None of these problems was unforced. They all involved missing data points in the knowledge base. In many cases, they resulted because the space network (and its properties) had not yet been discovered. The properties have been suspected for some time and have been ascribed to "dark energy."

Some may view this as shortsightedness in the discipline of physics. However, we (physicists) knew what we needed. We just could not find it.

There have always been many physicists that think the universe can be understood within a framework of ***random*** behavior. They have tried very hard for a very long time. If this were truly possible, it is likely they would have succeeded. This book has revealed a very large number of questions that cannot be explained by a random universe. The narratives of **MU** demonstrate how all of these questions ***can*** be answered within an ***ordered*** universe.

Who could have foreseen that gravitation was simply a byproduct of the warp fields of two objects "activating" each other when they get within range? This relationship was unsuspected since the force of gravitation causes acceleration, whereas the impulse that maintains a warp field only causes acceleration when it is increased. At first blush, this seems to exclude the equivalence principle from this discussion.

If millions of great minds had been stymied for a hundred years for lack of one or two discoveries, the world today would be a different place. Albert Einstein and his contemporaries demonstrated true genius when they invented a mathematical detour (universal c) around the trouble areas (like the space network) that allowed for extended developments in science until the physical knowledge gap could be filled. Universal c was the simplest solution available at the time. I view their achievements with great respect. Now what?

More like a Fable than a Fantasy

Long, long go, "Earth" had an "ethereal" space network called "Nettie." However, Nettie was a young, "spacy" piece-of-work. Earth's attention was soon drawn to "Constance C." and everyone thought that C would be our one true constant. We have now discovered that C is two-faced. Furthermore, even as Nettie has gained relevance through a total fabric make-over, we have gained wisdom; we are beginning to realize our mistake. The question now is, "Will we have the courage to admit our error?"

Acknowledgements

The extended development within this book can now be used to determine whether gap-zapping like this has correctly filled some of those potholes. Working on this road crew has been a great pleasure for me. Thank you for taking the time to consider my point of view and my life's great passion surpassed only by my wife Arleen, our six children, our ten grandchildren, numerous other family, friends and very helpful and encouraging colleagues and students, my proclivity for teaching, and my great love for God and Country.

I know, I know. That's a long list of exceptions, but in addition to that, special recognition is also well deserved by my late parents and my recently departed oldest son, Rick. Before his death, Rick was included in the six children mentioned above. His love for science and for me will memorialize him there for the duration of this book.

Thank you all for your support and understanding!

Max

10th Uni-Verse

"Photons uniformly raced,
One kilocog per second;
Flashing through the night of space,
As silent ticking beckoned."

11. Advanced Speculation

Introduction

Many of the hypotheses in **MU** were likely out of the mainstream of physics and therefore unacceptable to many theoretical physicists when, in the mid-20th century, they were first considered by a young scholar for inclusion in this eventual publication. Other speculation in this presentation is probably still beyond the reach of current research.

To reduce the probability that skepticism concerning the latter might taint the former, some of the less well-developed speculation that was chosen for inclusion has been quarantined to this back-of-the-book section called "***Advanced Speculation***," abbreviated, "A-Spec." The publication of this speculation might better serve to include other "advanced speculators" in conversation concerning this speculation in addition to corrections, comments, and admonitions concerning the ten concepts above.

Starting at the Beginning

The fundamental goal of physics has always been the unearthing of any and all knowledge concerning the behavior of the physical universe. The driving force has always seemed to be the desire to be able to predict what will physically happen in any particular natural setting well before it happens.

It was thought that this ability to make predictions would prove that the understanding was correct. After all, physicists had some early success in the predictions of eclipses. This did establish their understanding of those celestial patterns.

The ability to make accurate predictions might demonstrate understanding if the predictions were made on the basis of "how it works" instead of on the basis of derived formulas that are based on observed patterns. This is especially noticeable when the predictions are based primarily on computer models.

Some pattern recognition does sometimes require considerable mathematical skill; however, it does not ***necessarily*** require scientific insight.

Learning how to make accurate predictions based on observations of patterns without understanding the scientific behavior is a great mathematical shortcut, but it leaves us vulnerable to surprises.

The *study* and *use* of patterns constitute the *science* and the *art* of mathematics respectively.

Most mathematicians realize that different patterns can appear identical for a limited (but not necessarily short) time. Furthermore, most scientists realize that scientific patterns can change abruptly for sometimes obscure scientific reasons. This seems normal. "Fluctuations in patterns" also seem common.

Pattern Fluctuations

Patterns that are based on measurements can fluctuate due to round-off error. Patterns can often seem to fluctuate when the complete pattern is not well-understood.

However, pattern fluctuation can often be ascribed to a supposed randomness of natural behavior. With better attention to round-off error and pattern detail, we might begin to notice fewer claims of pattern fluctuations, especially when those patterns are supported by narratives.

One common problem is the confusion between cause and effect. Getting these reversed can make it impossible to recognize the pattern. One good strategy is to reverse the two whenever predictions are often incorrect.

For example, it is known that increased atmospheric temperature will allow increased carbon dioxide concentrations there (among others). Trying to use computer models to prove conversely that increases in carbon dioxide concentrations will cause increased atmospheric temperatures will likely lead to confusing conclusions.

For physics to have any credibility, it is not enough to know what should happen next. It is also necessary to know what <u>might</u> be an alternative and what factors can affect the outcome. This requires not only a knowledge of physics and an understanding of mathematics; it also requires an ***understanding*** of physics.

The problem with starting at the beginning is knowing where that is, exactly. For centuries, we have been working our way "inward" toward the fundamentals of all of physics.

The study of alchemy turned into chemistry during the "Age of Reasoning" of the 17th and 18th centuries. Since then scientists have progressed from the study of metals, to crystals, to molecules, to atoms, to particles, and now all the way toward an understanding of dark matter and dark energy.

It has been hoped that the most "basic" components of the physical world would have a simple structure that would be understandable. One or more of those components might lead us to a formulation of Stephen Hawking's "Theory of Everything." This then should tell us how "everything" is related.

We now have, and are developing many more pieces of the puzzle, but we still do not know how all (or even most) of it works. (See Chapter 1.) Much has been learned and the subject is stimulating, but progress toward real understanding is sometimes frustrating.

While this exciting work continues, is it not also possible to work from the other end? Might it be possible to learn more about how things work and how the pieces fit and work together even while we learn more about the pieces, themselves?

Might it not be admirable to strive for an "Understanding of Everything" without having to come up with yet another "Mathematical Theory" to help us explain everything? Maybe we can develop an overriding physical "***Narrative of Everything!***"

Narrative Physics

This branch of physics is being created to provide a standardized method for removing physical phenomenon from the mathematical realm of regality.

1. Choose a physical phenomenon for which you feel there is yet no acceptable *narrative*.

2. Make sure the phenomenon can be expressed physically regardless of any mathematical formulas.

3. Consider as many workable options for the physical assumptions as possible.

4. Choose the *narrative* that seems simplest.

5. Follow the implications.

6. If contradictions or impossibilities occur, alter or abandon the assumptions or the line of reasoning and pursue another.

7. If a line of implications leads to new implications or simpler understanding of "older" implications, the line of reasoning should be pursued further for additional proof or consequences.

8. One of the most exciting developments occurs when one simple implication unexpectedly taps into an elegant solution to one of the many baffling mysteries of physics. This progress will be called a "happy tap."

9. Substantial confirmation, or even one or two happy taps, might form the bedrock of inductive reasoning and can sometimes remove a physical phenomenon from the mathematical realm of regality. It is also important to develop adjacent relationships.

10. Some "senior" physicists might be reluctant to invest much of their remaining time on this new strategy if their priorities are already set. It might fall to younger physicists to pick up the mantle.

Casualties

As mentioned earlier, there are some current concepts of physics that will not survive the scrutiny of narrative physics. Most of these have been discussed at length. The (alphabetical) list includes the following.

- ✓ Equivalent mass
- ✓ Expectation of randomness where narratives might exist
- ✓ Length contraction
- ✓ Lorentz transformations of length and time
- ✓ Space warp
- ✓ Space-time
- ✓ Synchronicity loss
- ✓ The (literal) big bang
- ✓ Time dilation
- ✓ Universal c
- ✓ Universality of relative motion (with no absolutes)

What are the relevant questions?

Be prepared to admit that some concepts in physics were developed in mathematical cocoons with only gossamer connections to "reality." To investigate these concepts within reality based asies, it will first be necessary to transform the chosen asies into more relevant assumptions.

Do not give up; "the truth is out there!"

The questions we should strive to answer might follow a progression as follows.

a) What observed "behavior(s)" are we interested in studying?

b) What seems to be producing the observed behavior(s)?

c) **Never assume** that any behavior is random or even just natural!

d) How is each behavior related to its cause?

e) What known properties of the cause might motivate the behavior?

f) What is the observed relation between the cause and the effect?

g) If the relation above is not clear, we might speculate as to what relations are possible that *might* cause the observed behavior.

h) Once the mathematics of **d) - g)** is established, it is time to seek the narratives for the behavior; how does it work physically?

i) If a narrative cannot be found, circle back to **c)**.

The Clumping Problem and The Pooling Solution

The clumping "problem" relates to the lack of a viable *narrative* for the "strange" ability of galaxies to retain their densities despite the expanding universe. The clumping problem might be a *happy tap* into "A-Spec." Due to their great size, the structural cohesive force for galaxies is primarily "far-reaching" but relatively weak gravitation.

The force of gravitation increases linearly with increasing mass (proportionally to the mass) but it diminishes rapidly with distance (as the square of the distance).

The formation of galaxies plausibly requires pooling of dark matter to provide sufficient density (mass-per-unit-volume) for the gravitational intensity (energy per-unit-volume) necessary for structural integrity. Pooling refers to the condition caused by bunching particles together into "pools" while reducing both the internal "gaps" and the "spread."

The long-term survival of galaxies is also assumed to require pooling durability of dark matter. Consequently, we wonder how pooling works and exactly how that is related to the presence of singularities.

D-pushing

Due to singularity "bundling," the space network of fabrons (including the fabrons within galaxies) may be able to use "D-pushing" to maintain its density, while expanding.

Since warpons are probably not singularities then they are unable to bundle by D-pushing. Therefore, warpons within individual galaxies are effectively being forced to physically "pool together" to maintain their individual intensities.

This prevents galaxies from increasing in size (and thus decreasing in density) as the universe of fabrons expands around them. This seems to remove "galaxy pooling" from the realm of regality. This may even be how warpons were able to form independent protuberances within the early space network.

It is difficult to miss the obvious similarity of ***constant density*** of galaxies and ***constant density*** of the space network. To understand if and how these two phenomenon are related might require a knowledge of how motrons interact with fabrons. Specifically, we need to know how a galaxy is able to adjust to the singularity dole.

We now know that galaxies are contained within the space network. However, since the network is expanding as the galaxies pass through, how is galactic motion even possible unless their motron densities are *compatible* even as they continue on their separate ways? Since network density is maintained by D-pushing, then it may be possible that all galaxy densities are maintained at the same density as the space network by what we might call "compatibility pushing," or "C-pushing."

Before we go on our way, let us consider one other possible relation. Since the network is maintaining its density as it expands, this might help to prevent the galaxies from collapsing, or more likely evaporating behind it, as the network "expands" through it.

Spiral Galaxies

Logically, there would seem to be a limit on the size of spiral galaxies that could be maintained by pooling. This may be a very large number, but limited even as the universe expands.

Expanding the volume of a galaxy without changing its density would increase its diameter. This would then increase the centripetal force necessary to counteract the rending force by the linear bent of its rotation. Some might argue, "spiral galaxies only look like they are spinning, but that they are not." Strangely, this is a paradox within an illusion. Rotation seems to be the only viable solution to the gravitational collapse of spiral galaxies.

One of the complicating factors in counteracting the linear bent is that the force of gravitation diminishes as the square of the radius of the galaxy.

There may also be a minimum galaxy size. This, however, might be simply a matter of reclassification.

Maintaining Network Bent

The universal tendency for stable motron density over time *will be called the "network bent." M*aintaining the network bent is probably not optional. It is forced by photon pressure within the irresistibly *uniform* density of the expanding space network.

Note that photons may seem insignificant individually. However, innumerable photons of varying wavelength and strength are collectively traveling at very high speed in every possible direction as they continually and simultaneously "flood" virtually every cubic micron of space.

This is not immediately obvious, but it can be demonstrated. By moving and swiveling around as we observe, it is possible to realize that any object, barring an obstruction, is literally visible (with appropriate sensitivity and at the appropriate angle) from almost everywhere in space.

Therefore, barring obstructions, light beams from virtually every luminous object in space must be crisscrossing, collectively, through virtually every tiny point in space. Remember that this crisscrossing is occurring naturally and systematically within the space network. Except for the red shifting of universal expansion, the photon bombardment might be more destructive than constructive.

By this scenario, it may be possible that dark regions in space are less stable. This could serve to restrict their size and their durability. However, most dark regions in space are easily penetrated by non-visible light waves. ("Dark" generally refers to lack of penetrability of visible electromagnetic radiation.) Nonetheless, "stability" may be the key to pooling.

Constancy of Parochial C

There must be some reason that parochial c is constant. "A-Spec" will assume that D-pushing (of photon pressure) is responsible for motron uniformity and stability.

Motron density D seems to be the only parameter that might affect the value of parochial c. As long as D-pushing of photon pressure maintains the current value of D, then parochial c will be constant.

This *narrative* should remove "constancy of parochial c" from the realm of regality.

Furthermore, the fact of parochial c should establish network bent. This then allows us to simplify by replacing "Special Relativity" with "Natural Relativity."

Motron stability seems to solve the galaxy-clumping problem without the need for further dark matter. This would then solve the so-called "vacuum energy problem" and remove it from the realm of regality. If this solution is verified, it could be called the "singleton solution (to the pooling problem)."

This solution essentially replaces the concept of "vacuum energy" with "fabrons," since the energy of fabrons provides the "fabric" that subjugates the "vacuum" of the space network. Note, however, that (1) the term "vacuum" implies empty and this does not well-describe the space network and (2) the expression "vacuum energy" implies energy of a vacuum rather than an energy field within a mass-vacuum.

Both of these are just semantics, however. As long as the understanding is there, then fabrons might safely be called "vacuum energy."

The current definition of fabrons, however, allows that they also exist within warp fields (which contain mass). This would imply that "fabron" is a broader concept than "vacuum energy." If a better description (besides vacuum energy or dark energy) is needed for fabrons, "space energy" might work. After all, the name "fabron" was derived from the expression, "fabric of space."

Warpons travel with their associated object, and they provide its mass. This was explained in "***Mass Concept***." The confirmation of the singleton solution would also seem to confirm the contention that ***warpons*** hold the answer to the question concerning the abundance and location of dark matter.

Motron Energy

MU has presented a case (above) for the existence of a space network of dark energy called "fabrons" and migratory warp fields of dark matter called "warpons." These both consist of motrons of energy.

Convection and radiation through space are both achieved with the aid of motrons. Furthermore, "A-Spec" will assume that a warp field is a collection of q-pons, the carriers of "dark matter," including the not-so-dark photons. We will also assure that a fabron is a singularity of individual "q-fab" particles that collectively are stationary providers of dark energy to all massive objects and even to q-massive photons passing through the "bedrock" (the space network) of space.

Warpon Mass

This book has proposed a relationship between warpons and mass of all matter. Warpons also seem responsible for afterwarp and other warp fields.

Along with relationships discussed earlier, this current discussion seems to imply a relationship (probably equivalence) between the kinetic energy of a particle and it's mass even at the quantum level. These seem to supply the narrative for the relationship, $E = mc^2$.

It seems logical, therefore, that a ***photon*** and by extension *any* ***monopon***, having q-energy, must also have q-mass. For convenience, this might be considered to be the "equivalent mass" (using current terminology) of a light quantum.

If this could be established and the gram-equivalent of one q-mass could be determined, then the mass of any object could be expressed as an integral multiple of a warpon of unit mass. Number one on this list should be the photon with a mass of one warp unit (1 wu).

Since companion warpons escort all of their linked objects being transported through space, then through their common resistance to changes in motion, the companion warpons share their mass with their associated objects. This may be how photons and other forms of quantum energy receive their q-mass.

Fabrons are super-abundant and thus probably occur most often (or possibly always) in singularities. (This might be difficult to verify if it turned out that fabrons only existed in the space <u>between</u> galaxies. Please see the "Interchangeability of Fabrons and Warpons" paragraph below.)

During the Big Bang, the reduction of fabron singularities into more abundant singletons probably caused the cosmic expansion known as inflation. However, the companion process to inflation called matterfacation also produced matter. Thus, the conversion from singularities (fabrons) to singletons (warpons) was an integral part of matterfacation. Please see "***Cosmic Flatness Concept***" above.

The photon is <u>the fundamental carrier of electromagnetic energy</u> (via warpons). In this scenario, photons have been assumed massless beyond the q-mass of the carrier warpons.

However, it seems simplest to include the q-mass of a photon carrier with the mass of the photon itself. For all of the reasons, both stated and implied in **MU**, it is time to recommend that all monopons be considered to have q-mass of 1 WU and q-energy of c^2 $(m/s)^2 = 1$ $(kcg/s)^2$.

Many more facets of these relationship gems are yet to be hammered out by many "gem" cutters with sufficiently sensitive hammers. Invariably, some gems will likely turn out to be little more than dross.

Motron Stability Conjecture (Continued)

The process with the potential to maintain galactic stability is for all fabrons (or at least many fabrons) to be "bundles" of energy such as singularities that are capable of expanding as the universe expands (in a process called a "singularity dole"). This process allows the universal density "D" of fabrons in space to remain fixed regardless of cosmic expansion.

Notice that this type of (energy) expansion would <u>not</u> affect temperature since the energy involved would maintain constant density and hence constant pressure. This *narrative #19* should remove the concept of "isothermal expansion" from the <u>realm of regality</u>.

Isothermal expansion does not prevent the universe from cooling, however. Singularity reduction, itself, is a major drain of universal energy.

Intuitively, the inestimable fabron membership "list" must be constantly growing (probably exponentially) since fabron singularities must relentlessly be releasing fabrons to maintain constant motron density D on the parochial level. Using a singularity dole by fabron release to maintain the constancy of D is being called "D-pushing."

Note also that the singularity dole is also necessary to provide the additional network material necessary to fill the new turf requirement brought on by the very rapid expansion of-and-throughout the universe.

This singularity "dole" of a fabron singularity could be considered a systematic expansion of the volume being increased. To avoid confusion, this tendency of the space network to expand while maintaining density D is being called, the "network bent." This *narrative #20* should remove the concept of "network bent" from the realm of regality.

"Interchangeability" of Fabrons and Warpons

It has been speculated above that all fabrons are joined together (or at least they are working together) throughout the universe, including the space sub-networks within all warp fields, to form an entire space network of singularities. One alternative possibility is that each warp field is a closed, all-singleton sub-network of the singularity space network. This option might possibly be easier to understand and to explain than the option described in **MU**. However, it seems more complicated at this time.

For example, even if the space network were composed of fabrons only, warpons carrying photons and even warpons carrying whole galaxies would likely be persistent visitors as they travel through. How would this affect other objects traveling through?

Would the "converse need" for fabrons to coexist within warp fields presage further technical hitches? The over-riding question is, as the warpons transport their massive objects through the fabrons of the space network, are the fabrons simultaneously coexisting within the warpons "like clouds filtering through the air?" On the other hand, are they more coherent "like a baseball traveling through the air?"

Possible Link between Fabrons and Warpons

There is one intriguing consequence of warpons being "q-pons" instead of bosons. It might be that a warpon could simply be an elementary particle similar to a photon except that, after matterfacation, it carries a "mass quantum" (q-mass) that transitions (is absorbed) quickly (or simultaneously) among the fabron singletons (energy) of a particular object.

This might be similar to the way a light quantum (photon) travels through the fabrons of space! It seems to be required (upon matterfacation) for fabrons to give up their singularity status.

Under the q-pon definition, it is obvious that q-mass must be considered part of the photons that they transport. This q-mass would then logically replace what had been considered the mass equivalent of the light quantum.

We know that mass and energy are equivalent but different aspects of equality. Q-mass and q-energy seem quite different, but it is likely that q-energy = q-mass · c^2. The physical difference between fabrons and warpons might provide a *narrative* for a better understanding of mass. See "***Mass Concept***" above.

These complications seem best left for future development by continuing to assume that fabrons are distributed uniformly (with the same density) through all warp fields.

Causing Absolute Motion

The existence of a space network of fabrons has now been well established by MH. This has then established the possibility of absolute parochial motion. This was discussed in Motron Hypothesis above. Until now, it has been thought that all motion was relative. The discovery of warp-force will bring many new adjustments in our thinking.

We have long known that unbalanced force causes acceleration. We must now analogously decide what causes momentum. This concept is already understood, but we do not have a name or a narrative for it. Since force causes acceleration and acceleration causes speed that causes and increases momentum, then transitivity of this "causation" implies that force also causes momentum.

There is a difference though; it seems pedestrian to quibble about whether the increase in motion has direction such as positive, negative, and zero. However, when any action causes any motion, it now matters.

The simplest solution, at least for now, is to consider all motion, not just acceleration to be nonnegative and caused by "force."

One type of force causes acceleration. This is the first type of "force." The second type of force, with passing time, causes non-increasing speed such as momentum. This "warp-force" is not the same as current "impulse." Impulse lasts for a specific length of time during which the object is accelerating. Then what?

The momentum continues but there may no longer be any force or impulse. Since it is unbalanced force that causes acceleration, then warp-force might be considered a "balanced force."

Now, momentum does not just happen. It is caused by force and maintained by warp-force. However, this is a fine point that may be only semantical.

Q: Why do we need to change? Why not just continue to call them force and impulse, respectively, as we have always done?

A: The recognition of warp-force will cause more adjustments than just this one.

For one thing, there are four fundamental forces. Three of them only cause acceleration. We now see that warp-<u>force</u> can also causes momentum. This was not noticeable when all motion was thought to be relative. It would very likely be awkward to continue saying, "Force causes acceleration."

Instead of trying to fix that, it might be easier to say, "Force causes motion." Newton's three laws will then be fairly easy to fix.

We will begin this discussion with a restatement of one of the most basic discoveries of all of science.

The Narrative Principle

All physical phenomenons occur for a physical reason that can be expressed as a "narrative."

Seven (Revised) Laws of Motion

<u>*1st law*</u>: The space network (although expanding) is always assumed stationary within (as) the uniformly expanding universe.

2nd law: Absolute motion (with respect to the space network) is "parochial" (local) motion. All motion not referenced by frame must be assumed absolute.

3rd law: All motion with respect to a moving frame is "alien" (not parochial) motion.

4th law: All (linear, angular, and orbitar) momentum of every object will be maintained (constant) by the corresponding warp fields unless some interaction (force) changes the object's state of motion.

5th law: Acceleration a, of every object, is directly proportional to the corresponding net effective force f_e and inversely proportional to the object's (rest) mass m_0: $f_e = m_0 \cdot a$.

For angular and orbitar motion, centripetal acceleration a_c, of every object, is directly proportional to the corresponding net effective force and inversely proportional to the object's (rest) mass m_0: $f_e = m_0 \cdot a_c$, where $a_c = v^2/r$ and r is the radius of curvature of the motion.

6th law: The composite (alien) motions of all three types of speed and acceleration are the (corresponding) vector sums of all corresponding component motions of and within all active frames.

7th law: To every action of every type on any object there is an equivalent and opposite reaction of some type, depending on the construction of the object and the constraints upon the motion of the object.

It is likely that these seven laws will improve considerably with time, as inward simplicity and outward breadth are balanced and enhanced.

Parochial Momentum

The 2nd law of motion is guaranteed by:

a) Parochial c is constant,

b) Mass ($m = m_0$) is independent of speed (v), and

c) Speed (v) is constant whenever $a = 0$.

Therefore momentum $p = mv$ must be constant whenever $a = 0$ and m is fixed.

Origin of Space

Now that we know about the space network, it might be appropriate, without going too far afield, to speculate about the origin of the universe.

Within the formal Big Bang theory, there are many formulas, but no real narratives. Everything is considered to have occurred randomly. However, since the Big Bang theory implies a definite beginning and a probable end to the universe, then the possibility of order in the interim must be investigated. Overwhelming evidence seems to confirm this premise.

It is likely that long before the currently acknowledged creation process, cosmic temperature and pressure were building. Yet, lacking a space network at that time, there was no means of travel available to expand and disperse the temperature or the pressure. The pressure to expand and the inability to do so caused the energy to remain where created and accumulate into one or more singularities.

MU will assume one location and call the "original origin," verbally "O-naught" and symbolically, "O_0." Since the universe did not yet exist, O_0 might be called the original "middle of nowhere."

Scientifically, we can only surmise that by some process (currently under inquiry), temperature and pressure eventually became great enough in this one locations for the formation of all of the now existing elementary particles from energy.

The singularity of heat energy that accumulated at O_0 will be called, the "heat pool" even if that name might be contraindicative of a singularity. It might even now be possible to approximate the quantity of heat that accumulated in the heat pool prior to the Big Bang. Before we can continue, however, there are two eventualities that must be considered.

- Not knowing where the heat came from leaves open the possibility that heat could also escape the same way (in reverse).
- Under the conditions that exist today, energy can neither be created nor destroyed.

Since there is no evidence to the contrary, then it seems best to assume (for now) that all of the energy of the Big Bang is the exact same energy that is still present in the universe today.

The singularity temperature was thus able to climb to unimaginable heights before the epoch of inflation and ultimately to disperse with no "Big Bang-causing" shock wave. With no "bang," there was no explosion and no shock wave. It seems that the only similarity to an explosion was the hot expansion. Rather than continuing to perpetuate this misconception, it seems time to choose a new name for the beginning of time. For the remainder of this book this name will be abbreviated, CE for "creation event."

The following conjecture will be presented as speculation without motivation. In this way, we can avoid discussing the many other options that must be rejected for many other reasons. Once this conjecture is understood, we can establish its tolerability by studying the <u>many</u> positive ramifications.

Many people wonder, "If the universe has been expanding since the CE, where might the relic heat of the original CE now be located?" There is only one answer to this question, "The heat that had accumulated at O_0 before the expansion called "inflation" is now everywhere; it is still the entire universe" just as it was before and during inflation.

The 2.725 K remnant of that heat from inflation is today called the "cosmic microwave background radiation" or "relic radiation." This trace amount of heat permeates the universe.

It has not "traveled to the ends of the universe." it has remained (probably circulating) in place (at the now inflated O_0) while becoming diluted by the expansion of the universe. It is also assumed that no heat has been lost. (Where could it go?) The temperature, however, has dropped due to the dispersion effect of the expansion. The energy's travel has been limited because heat always travels from warmer areas to cooler areas. This makes accumulation of heat difficult. The heat of creation is, and (with minor fluctuations over time) always has been everywhere comparable other than isolated hot spots such as galaxies. Galaxies are also very cold with isolated hot spots such as stars.

MOAS

This discussion seems troubling at first since the original CE expansion is thought to have sprung from a singularity. It is nonetheless possible that this realization explains a lot about the universe.

It has been generally assumed that when a singularity expands that it is no longer a singularity. However, there may be other options for the super-singularity of the CE. In this case, we might call it "the mother of all singularities" or "MOAS" for short.

Since fabrons are particles of energy, then it is likely that most fabrons, or possibly every fabron, is always a singularity or at least a fabron "bundle." This would make the universe (MOAS) a singularity of singularities, like a "network of fabron factories."

Abundance of Dark Energy

In this regard, however, it is likely that each fabron singularity acts as a single fabron similar to a single photon, but without the q-mass that is associated with photons. Another intriguing verification is that the reason fabrons are stationary may be the possibility that fabrons have no mass at all. Whether hypothetical or actual, a particles of this type might be called, a "null-pon" or a "P_{ϕ}."

Eventually solving this mystery might also help ease the so called "vacuum catastrophe." For now, we will continue to assume that the simplest solution is for every fabron singularity to be a P_{ϕ}.

We have previously discussed the formula, $E = mv^2$. This formula clearly only applies to kinetic energy E since a value of $v = 0$ implies $E = 0$. Therefore, the stationary fabrons of the space network may still have energy even if they have no mass. Remember that we have already shelved the concept of "equivalent mass" except for illustrations. This makes fabrons invisible; they are thus truly "dark" energy.

In this scenario, the energy corresponding to the total null-pons within one fabron singularity could be many times as great as the total energy of a warpon singleton. Since null-pons provide the energy *of* the space network, then they are less noticeable since null-pons do not travel *through* the space network. This huge fabron energy might explain why dark energy ostensibly comprises 73% of the total density of the universe. This narrative might serve to remove "dark energy" from the "realm of darkness." This is significant since dark energy is considered by many to be the foremost enigma in astrophysics today.

If fabrons are so plentiful, then why has a q-fab particle not yet been found in space?

1. Singularities with high energy are difficult to manipulate.
2. Fabrons cannot be accelerated since fabrons do not generally travel.
3. Fabrons cannot travel through the space network since they are the space network.
4. It does not seem likely that the space network can be bombarded with massive particles because the space network channels the motions of all massive particles.
5. A singularity under stress might simply subdivide.

6. Fabrons in space may generally exist as high-energy singularities and may be difficult to locate in the q-fab state.

7. This search will clearly require considerable ingenuity.

Just as a light beam contains a collection of photons that each contains a light quantum, it may also be possible that a fabron singularity contains a collection of (q-fab) fabron singleton that each contains one or more null-pons. Since a physical fabron (probably due to its location and structure) has yet to be discovered, there seems to be a lot we still do not know about this most basic and most abundant of all “particles.”

It is important to admit at this point that if the space network did not exist, then virtually every concept in this book will fall apart. In fact, every theory in physics dealing with some version of space “fabric” would be thrown into turmoil. Thus, there is a lot riding on this outcome.

However, because there is so much confluence surrounding this idea, it is very difficult to imagine that it has not already been established as valid. A-Spec will thus assume this to be the case.

D-pushing Consequences

Due to the “control structure” that is palpable in the entire universe, it is likely that some process controls the number of individual singularities within MOAS and the contents of each.

The strongest heart requirement suggests that A-Spec at least temporarily assume that all fabron in the universe are generally singularities. As cosmic fabrons multiply by subdividing, their density within the universe is maintained by D-pushing stability.

This implies that D-pushing might be the process that maintains uniform distribution of the density D throughout the universe. It then seems likely that the number of singletons (null-pons) contained in each singularity is determined by the densities of its surrounding singularities.

For example, if one singularity has a slightly "high density," it might dole a q-fab) that might, in turn, cause a "less dense" singularity "near-by" (but not necessarily "close") to absorb an extra q-fab, etc. In other words, "one q-fab pops out here, another is absorbed over there, or sometimes not." This "whack-a-mole concept, on a universal scale, might maintain all singularities at or near the average (constant) "density," even as the universe expands.

Please note that it is not necessary for the doled q-fab to travel to the receiving singularity. Each q-fab is basically just a fabron singleton.

The entire space network is filled with fabron singularities accompanied by null-pons popping in and out seemingly at random (but probably not) "all over the place." We will now add the assumption that all singularities are maintained at or near average density most of the time, which means all of the time on average.

We must also consider the possibility that, from time to time, "seismic" events might happen to cause the total density of fabrons to excessively increase (or decrease) causing the content of many of them to decrease (or increase) as compensation. This would likely cause a "wave" of adjustment to ripple through the universe at light speed. We could call it a "null-pon" wave.

Multiplying by Subdividing

Another alternative to total singularity reduction and the singularity dole is reduction by sub-division. This refers to the process of sub-dividing every fabron into two similar "half-fabrons." This possibility seems to correlate with the ***inflation*** result of doubling the volume continuously over time.

One possible process to achieve this goal would be for each individual fabron to subdivide into individual fabrons that immediately begin sharing surplus null-pons with adjacent "subpar" fabrons. This process would differ somewhat from the dole process in that q-fab would pass directly from one fabron to another without first popping out. This version of reduction would result in greater variety in q-fab diversity. Since all that really matters, however, is the average q-fab mass, then the sub-division model should work well.

The Story

The following discussion is extremely speculative, but it is one possible scenario for the CE. The story begins with a probably infinite void we will call "space" and a probably finite accumulation D_ϕ of amorphous plasma of dark energy in a place called the "heat pool." Initially lacking a space network, the energy could not disperse through space.

With no space network, there could be no dimensions. (There will be a distinction between a D_0 singularity point and a non-dimensional, "amorphous" space "D_ϕ".)

Transition to Order

We have been discussing many possible procedures and processes that likely worked together to produce the universe as we know it. Many of the actions were caused by other actions, which caused other actions, etc.

By studying the cause and effect of each action, it might be possible to determine a potential sequence for the creation events. Much about these events is already known or suspected. Many of the details of the following list are contained in this book.

There are many possible sources for the primordial heat that existed at the beginning of creation. This heat is generally called "dark energy," but in this book, these particles of energy are being called "fabrons." Fabron singularities almost certainly did not come from any supposed massive explosion that might be called a "Big Bang!"

Possible Sequence of Events

1. We will list heat as the "first cause." This is the step that some call the "Big Bang." Regardless of its name, it is unknown if this accrual of energy was swift or if it was amassed over unimagined eons before the official beginning of time. Thus, in a pre-scientific way, the "first cause" is more easily and commonly thought of as God. How could that not be acceptable? We are not talking about science. We are talking about the inexplicable! The "first cause" can be comfortably left within the realm of regality.

2. Heat energy provided an elementary process (leading to the "creation") we will introduce shortly as "fabrication."

3. Fabrication probably produced the earliest elementary particles including photons and motron singletons.
4. Photons would soon provide us with photon pressure.
5. Motrons would then provide the space network.
6. Photon pressure would provide D-pushing.
7. The space network made motion possible.
8. Motion and D-pushing resulted in universal flatness.
9. Universal flatness was likely responsible for the universal stability provided by the network bent.
10. The network bent was likely responsible for singularity reduction.
11. Singularity reduction with D-pushing made matterfacation possible.
12. Matterfacation produced inhomogeneities and protuberances.
13. Photon pressure and universal flatness were likely responsible for inflation that likely ended matterfacation. The space network and singularity reduction were also necessary for inflation.
14. Matter, motion, and SNA made the universe what it is today.

Details of Creation

Inflation and matterfacation would have been natural consequences as the fledgling universe increasingly became heated. However, it is still important to seek to determine the sequence of events that may have caused inflation and matterfacation to occur.

Singularity Doles

MOAS, while maintaining its own fabron super-singularity, eventually doled out some of its singularities. However, this was probably not so much a "consequence" driven by MOAS, as it was a bizarre breakout "escape" by some of the fabron "inmates."

Since only energy can exist in singularities, these "singularity doles" allowed some of the doled fabrons eventually to fabricate some of their very hot, compressed energy into elementary particles, (including photons, motrons, and null-pons). Thus, singularity doles resulted in sequential "loss of singularity status" for some of the affected fabrons.

This seems to imply that many altered fabron singularities were converted into numerous warpon singletons that are now all associated with specific massive objects. (Details to follow.) In this way, every massive object is associated with particular warpons. The unaltered singularities, however, remain as fabron singularities of the space network.

It might turn out that the **singularity dole (in concert with D-pushing) was primarily the responsibility of matterfacation of the universe**.

If established, matterfacation could remove the singularity dole from the realm of REGALITY. Much more speculation will continue next.

The singularity dole, matterfacation, and inflation all required large expenditures of energy resources. Due to the time constraints on chemical reactions, matterfacation was the key process that required significant time to complete.

Given enough time, the matterfacation may have been able to eliminate the over-heating and the extreme pressure. However, the singularity reduction that was necessary for matterfacation momentarily <u>increased</u> the photon pressure. This may have been the primary "trigger" of inflation. Furthermore, there is one other criterion for pressure.

Since most of the pressure was in the form of radiation, then there could be no inflation until the availability of sufficient radiation. Keep in mind that D-pushing also requires photon pressure.

We have discussed the fact that photons not only have q-energy, but also q-mass. This implies that inflation cannot begin until matterfacation is at least initiated.

One of the initial step of matterfacation involves the conversion of energy into elementary particles. We will call the likely conversion process, "fabrication." No lie!

Polarization of the null-pons would discriminate one dimension of the unit. This unique dimension will be called, "mini." Due to the "<u>growing</u>" process and the confinement of the extreme heat to a specific "pool" area, this dimension, mini, would have been relatively "thin." The remaining <u>two-dimensional</u> subspaces of the unit would have been cross-sections of the fabrons of energy that will, as they are individually mattified, be called "<u>bilarities</u>."

The bilarities can be thought of as very large, thin D2 sheets of energy particles that were being formed with minimal (mini) thickness as the unit of bilarities "grew" like a (not-so-short) "short stack" … one "flapjack" at a time.

Due to the lack of gravitation at this early stage, this short stack might just-as-well have been horizontal (as if there was any difference before gravity).

Initially, the short stack was not expanding (due to the lack of a space network). However, space must have already been either a D3 space, in the process of becoming a D3 space, or amenable to a D3 subspace because the MOAS singularity reduction was pushing the D2 short stack rapidly into the third "mini-dimension" with each succeeding singularity reduction.

To continue our accumulation of metaphors, each "sheet" of the short stack "ream" became a very short wavelength, broad-beamed "mini-thin" sheet of null-pons that are, in our story, now being converted, by matterfacation into "bilarities." The short stack of bilarities will be christened the "Earth block" due to the earthy promise of its existence. It is significant that the original Earth block consisted of null-pons of energy and was void of matter (or even structure, initially).

This Earth block grew very rapidly, but only one sheet at a time as the requisite heat pool remained fixed (where it had always been) while the Earth block stack of bilarities grew (literally) from it. This process will be called "sequential matterfacation."

Particle Physics

The conversion of energy into particles of matter has been well reasoned and well documented in an area of study called particle physics. Thus, there is no need to reproduce that reasoning on these pages. Furthermore, the terminology in this book has mostly been developed independently of other studies.

Author's Note

The vision that is being developed in this book has been developing for decades. Yet the vision I have has still not been completely verbalized. Certainly, I am not the only one with this vision. There being no end of possibilities in sight, it is time to conclude this portion and get it to the publishers and begin the process of reconciliation.

To prepare this message for the transition to other voices, let me see if I can articulate what I have been privileged to see.

The Big Picture

Metaphors and similes are very important in this articulation. We are dealing with a development for which detailed science is lacking (or unfamiliar to many) and this causes a shortage of vocabulary to describe the action.

If one picture is worth a thousand words, then this might be a good time to paint word pictures to communicate the physical concepts. In other words, for this discussion we will be subjugating our predilection for "how it works" in favor of "how it might have looked."

In this "picture," do not worry about our perspective. We are not viewing from within the picture; we are simply "gazing upon it."

Let's start with what we have <u>not</u> been able to see. What we have labeled the "heat pool" was probably an extremely large quantity of energy in an extremely small space.

The heat pool, itself, is not likely visible. Only the earth block emerging from the distant heat pool is likely evident as it "streaks" across the field of view like a meteor.

For one thing, there may have been nothing to see until the advent of the space network, which is required for even light to travel. For another thing, the heat pool would likely have been shrouded by very thick "clouds" of cooler energy.

However, we may now plausibly speculate what the Earth block may have "looked" like. In our mind's eye (or "vision"), we are "witnessing" the birth of the space network. Photons and motrons were probably among the first things to be mattified. Yet lacking a conduit, energy (including light) could not travel.

Expectation

Due to its extremely small size and great distance, the energy pool would have been difficult to see. If the energy pool were a singularity, it would have been virtually invisible.

The great size and distance, necessary to make sense of this description, would make the motion of the Earth block seem much slower than it actually would have been.

As the apparent white-hot "core" of the Earth block streaked across the fledgling universe, it would have appeared to us like a very bright, rather rapid meteor leaving a very long trail in the black sky.

We would learn that what looked like a trail was actually the "trailing" portion of the actual Earth block itself, not just some vapor trail. The "core" was the leading face of the short stack. Earth block was not <u>traveling</u> through space but <u>into</u> "pre-space." It was destined to <u>become</u> the space network.

Due to symmetry consideration, the Earth block would probably have had a rectangular cross-section. Thus, but for the shroud and the intervening distance, it likely would have resembled a <u>very</u> long <u>rectangular prism</u> as its leading face (the original bilarity) surreptitiously "roiled" like lightning before the thunder, silently through the darkness.

However, the Earth block may have started out as a very thin streak but soon began growing thicker even as it was rapidly growing longer. It may soon have <u>seemed</u> destined to become a prism rather than a sphere. We can now reason, however, that the "parallel" dimension of the "eventual sphere" developed more quickly than the other two dimensions.

Possessing as yet neither substance, beyond its newly forming elementary particles, nor a space network, Earth block would provide no place to be or to go and no way to get there … yet. Due to its extreme brightness, and great distance from us, we would have been unable to observe any of the Earth block's structure behind the shroud, until it slowly and silently came to a halt. With no network, there was no momentum, but also no way to see anything with human eyes.

Even as the details of the bilarity become more distinct, it would still be difficult to tell whether the bilarity was approaching us laterally or if it was expanding unilaterally.

We now suspect that the "swelling" of the bilarity may have been due to lateral inflation. There is now some question concerning the timing of inflation.

We were either seeing it in slow motion, we were seeing it from a great distance, or else the expansion time has been miscalculated. This only seems a possibility if the calculations were based on possibly incorrect understanding (sequential matterfacation for example) of the actual creation process. Most will assume that the fault lies somewhere within the vision outlined above. Um, could be.

While arrayed in slices like a loaf of mini-thin bilarities, the Earth block, by its glow, would have made itself visible. However, as it cooled it is likely that darkness would have reclaimed the universe.

Distance can often be judged by the apparent size. If we would like to see the entire universe during inflation, then we would need to be very far away. Even a very great distance, however, would only cause "sequential matterfacation" to be visible as an apparent "meteor" if the process took several seconds.

For what it's worth, it seems that sequential matterfacation, within this "sequential matterfacation" version of the CE, would be much slower and last much longer than would "bubble" expansion of inflation. Keep in mind that the horizon may not be a problem if the universe is narratively driven instead of being totally or even partially random.

The greatest difficulties in gauging the timing of these processes are knowing the likely speeds of fabrication and matterfacation and the number of individual separate steps in the fabrication process that created the individual bilarities. More importantly, what was the precise sequence of CE's? (Progress by particle physicists toward this understanding has been truly breathtaking.)

This fledgling prism, having become fabricated, should still be called "Earth block." Though Earth block was not the original universe, it was, instead, the birthplace of the early building blocks of the universe.

By this time, at the dawn of the universe, the portion of the extreme heat of MOAS that would have been available to Earth block may have been severely drained by MOAS's singularity dole. In this state of development, Earth block may have had only sufficient heat for sequential matterfacation. Earth block, however, would have been (new metaphor) awash in a stagnant sea of multitudes of very hot singularities (similar to the fabrons of Earth block before their reduction) that would have likewise been released by MOAS's dole.

Every hot singularity of MOAS adjacent to a bilarity singleton would then have, sequentially, combined in various configurations to form a variety of elementary particles and provided them, by conduction, sufficient heat for their matterfacations as bilarities. The details of this matterfacation will be left to highly qualified particle physicists. The discussion will continue here with broad strokes.

"The MOAS Conjecture"

Regardless of the structure, the singularities of the space network will continue to be called fabrons. The previous paragraphs have raised many questions. One of the biggest puzzles is, "How might this all have come about?"

One problem with the facilitation of this scenario is that since the space network is responsible for cosmic expansion, then the space network must first exist before it can expand.

The swift depletion of the seemingly lavish supply of heat mentioned earlier seems to have placed constraints on continued expansion associated with inflation. Both matterfacation and singularity reduction (expansion) on a colossal scale, have placed high demands on the energy resources needed for inflation.

Since matterfacation only occurs at very high temperatures, then with time, singularity reduction could, and inflation would have reduced the cosmic temperature sufficiently to prevent matterfacation. However, matterfacation requires singularity reduction since singularities cannot contain matter.

Moreover, matterfacation is unlikely to occur after inflation due to the dilution (of the primordial heat) caused by the expansion called "inflation." This circular demand would seem to call for some type of jumpstart (sometimes called a "bootstrap"). This term comes from the nonsensical but well-warn notion of lifting yourself by pulling up on your own bootstraps.

The Bootstrap

The creation of a heavy concentration of quantum energy such as photons and fabrons likely produced the first moment of stability (to be discussed shortly), which would have caused an extreme expansion of the new fabrons to reestablish the universal density of D. This accomplishment may have marked the beginning of the epoch of inflation. Neither inflation nor cosmic acceleration likely requires the extremely high energy levels that were required for matterfacation.

Due to the high concentration of fabrons from the singularity dole of the bilarity, there was no need to acquire more by mergers and acquisitions.

Traditional cosmic structures, such as stars and galaxies could not form at this time due to two serious obstacles. There were no structures to concentrate gravitation and no gravitation to hold structures together.

It must be assumed that the earliest structures must therefore have been held together by singularity bundling, nuclear forces, or even exotic primordial forces that may have existed.

However, inflation was less of a rending pull and more of an outward compressive pushing force. Therefore, the reactive compression to inflation's expansion helps to control universal density. This compression is being called, d-pushing." It is most likely that, before inflation, there were no cohesive material structures.

Due to the probability of polarization and coherent space of the just released fabrons of energy, it is possible that every slab of fabrons that comprised the first energy structure was mattified quickly into literal building "blocks" of matter. It is also likely that the available heat required for matterfacation became available sequentially creating one layer of blocks after another over a very short period of time.

With all of the uncertainty, there are many possible scenarios for the first structure(s) of the universe; one possibility is (a) one or more giant spheres and another is (b) one or more gigantic prisms (probably rectangular).

In any case, the seething raw material became available just as the first block of elementary particles was cooling down from its literally “brilliant red hot” status and becoming visible through the concealing white “clouds.” It is important to understand that a cool down would not occur over time unless there was some energy drain such as inflation or matterfacation. This cool down was probably an indication of the establishment of the space network.

Option (a) for Earth block seems more problematic than the second option because of the lack of gravitation before warpons were established. Since none of these preliminary structures survived subsequent interaction, there may currently be scant evidence to discriminate the actual sequence of events.

“A-Spec” will speculate here that the earliest cosmic structure was one giant energy prism (the Earth block) that came into existence due to a temporary confluence of circumstances and quickly grew in length, within a glowing, white-hot opaque cloud of fabrons, changing to brilliant red, as it cooled, due to expansion and matterfacation. Until the space network became functional, cooling could not occur by “transfer of heat.”

Cooler white clouds may have formed there, surrounding the Earth block as it streaked through space. It is also possible that the clouds were pilfered from the energy cloud as the matterfacation continued within the earth block even after traveling beyond the boundary of the energy cloud.

After inflation, but still lacking massive structures, individual sub-blocks of energy began floating away from their Earth block of origin. (Remember, there was still no gravitation.)

This separation began at the “leading” end of the Earth block and spread quickly as each sub-block matured. Thus, the sub-blocks of energy began melting away faster and cleaner than ice cubes on a south Texas sidewalk in July.

This marks the end of the beginning and the beginning of time.

"Fabrication," "Growth," and "Matterfacation"

The creation process likely resulted in the current space network, which then expanded during the epoch of inflation.

Immediately preceding inflation, the stage was set for matterfacation in a way that minimized the horizon problem since every singularity of the MOAS was virtually the same age and "born" of the same circumstances. Moreover, they were all born in the same heat pool from the same mold. This eventuality, along with the general lack of randomness in the CE might help to explain the absence of antimatter in the universe.

This point of origin at the very beginning of inflation will be noted as O_I. The creation process to this point will be called "The MOAS Solution."

Beginning at this same initial time of O_I, The expansion of the space network, in addition to causing inflation, simultaneously allowed matterfacation to commence.

Before inflation, the universe was still tiny. However, inflation ended very quickly and after inflation, there would have likely been inadequate energy for matterfacation. Thus, the onset of matterfacation with its immense energy drain may have delayed inflation just long enough for the completion of any matterfacation that <u>was</u> able to occur.

The bilarity, having been reheated had become a very hot singularity of singletons. The immediate matterfacation of Earth block began the process of creating new elementary particles, including photons and motrons.

All of the currently existing elementary particles were swiftly produced in this process, likely still within the first minute (or so) of time. Some of the emerging elementary particles must have been what we now call "fermions."

Every fermion must have included an appropriate quantity of warpons, which determined it's mass. It is significant for warpons and fermions to be created early in the creation process so that fermions and eventually matter can be distributed throughout the initial universe by inflation.

Inflation

Here are some of the details as to how inflation must have played out. Before matterfacation was possible, MOAS was composed of a huge number of fabrons. With the singularity dole of the bilarity, MOAS gave up its own singularity status. Instead of being a ***singularity*** of singularities, MOAS became a ***universe*** of *mostly* fabron singularities. Earth block, then a collection of bilarities was "primed" for inflation.

It now appears that by giving up its singularity status, MOAS not only released the Earth block, but also innumerable fabrons while wielding the tool of D-pushing. This likely caused the sudden cosmic expansion called the epoch of inflation. For simplicity, we will assume the singularity reduction of MOAS into individual fabron singularities was totally responsible for inflation. This narrative removes inflation from the realm of regality.

The Horizon Problem

This "problem" deals with the fact that the entire universe developed "similarly" despite being out of communication range. Before trying to find a solution, let's consider whether this really is a problem.

As substances and systems develop, they gain more options and more variability as they become more complex. This can lead to greater range and greater diversity. This partially explains the biodiversity on earth.

The ***early*** universe, however, was relatively simple and still operating on basic principles. Furthermore, the narrative for inflation implies that inflation is globaconic. This then provides an obvious solution to this "horizon problem." Narratives are very powerful. They not only tell us how things work, they also tell us "why" they do not work some "other" way under the same circumstances. Thus, the reduction of the horizon problem does not eliminate, but does tend to alleviate the urgency for a speedy CE.

We should not forget, however, that the "horizon problem" is based on a negative. It provides no evidence that "things could have developed differently than they did." It only gives us one possible reason why they did not. The publication in **MU** of so many narratives for so many processes has reduced this expectation of randomness in the CE.

Inflation and matterfacation mark the momentous beginning of our understanding of the creation process. We could call this threshold at the beginning of time, "heaven's gate."

Inventory at the beginning of time

- Location
- Null space
- Darkness
- The Earth block

Evidence of the MOAS Solution

It is unlikely that the expanding of the universe into a bilarity before subsequently expanding into the current space network, as envisioned in the MOAS conjecture, could have occurred without leaving a trace. If ever discovered, this "fossil" of the MOAS "dinosaur" would carry the data search for universal genesis one full step backward into the darkness of the past. If such an anomaly is ever discovered, it may verify this conjecture.

The MOAS problem

Inflation, alone, may not have been sufficient to solve the horizon problem. **MU** is assuming that one or more bilarities of energy were converted quickly (during the first minute of creation) into elementary particles that were more complex.

We are also told that the universe (MOAS) expanded very quickly during the era of inflation. These two processes must have partially occurred simultaneously.

Once started, it might very quickly have converted, in a multistage process, every singularity into elementary particles. Yet, we also know that most fabrons survived (probably as singularities).

With the advent of the network and with the universe still transparent, photons, once created began to flood space. This likely resulted in constant density of the space network for the first time.

With constant density of cosmic energy, it is likely that matterfacation began in every available singularity of the relatively small early universe almost simultaneously. This conversion process must have instantly absorbed massive quantities of heat everywhere as did the massive expansion of the singularities into individual singletons.

Matterfacation from energy to elementary particles of greater complexity may have ended after about one minute (or maybe a little longer in the case of "sequential matterfacation") when either the temperature fell below the required threshold or every bilarity was consumed, or both (not necessarily simultaneously).

The conclusion of matterfacation will be called the "first moment of stability." This would have caused an extreme expansion of the new fabrons to reestablish the universal density of D. This accomplishment may have marked the beginning of the epoch of inflation.

Summary of D-pushing and Network Expansion

MU has postulated many controls over the expansion of the space network of fabrons. By all accounts, space, as we know it could not and therefore did not exist before the advent of fabrons.

During the epoch of inflation, electromagnetic energy traveling through the developing network created pressure on the fabrons as a byproduct of its predilection to straighten any curvature of space. Fabrons, in response to photon pressure, expanded rapidly to accommodate the need for flatness. When dynamic flatness was achieved, the motivation was reduced to maintenance status.

As the bourgeoning new space network "exploded" outward causing inflationary expansion, it simultaneously caused an overwhelming opposite (inward) reaction pressure on all motrons. This monstrous ***reaction pressure*** served to compress the fabrons to their maximum density (D) in a process we have been calling D-pushing. If D-pushing was not sufficient also to compress all warpons to the same density, then certainly C-pushing would have completed the job. (Please see "compatibility pushing," in ***Expansion Concept***.)

The minimum distance between fabrons of the space network and also between warpons of all galaxies is probably the Planck length $\ell_P = 1.616\ 199(97) \times 10^{-35}$ m. Therefore, the most logical value for D is probably $D \approx 2.368\ 728\ 52 \times 10^{86}$ motrons per μm^3 (cubic micron). This could be acting as a pressure gauge that continues to keep the density D of the universe constant. This could explain how the universe can expand and the temperature can drop, yet the density D not be diminished.

Finding the strongest heart is exciting science, but it does not imply the availability of a simple explanation. This expands, enriches, and braces the old Einstein quote …

11th Uni-Verse

"Everything should be made
As simple as possible,
But not simpler."
– A. Einstein

Addendum

Since the structure of this book was first laid out, there have been many new developments in the field of physics. As a result of these developments and of new insight by the author, some of the ***Advanced Speculation*** of this last chapter has found considerable reinforcement. These topics deserve promotion to the main text. However, it would not be optimal to hold up the already long delayed publication of this book for relatively minor updates. It may be time for the author to begin enjoying some of the benefits of retirement. Therefore, you may find some of the speculation in this last chapter to be less speculative.

Glossary of New Terms and Concepts

(In addition to some important Basic Concepts from Physics

1. New (<u>Y</u>es?)

2. Term_____________________________________

3. Meaning___________________________________

. . . .

Y **Aces**

The pronunciation of "asys"

Aether (or ether)

An earlier version of a hypothetical space network

Y **Afterwarp**

An adjusted warp field caused by (and maintaining) linear motion

Y **Afterwarp Equivalence**

The (linear) equivalence principle

Airfoil

Refers to the shape of an airplane wing

Y **Alien c**

The assumed (non-parochial) constant light speed

Y **Alien frame**

Any (remote) frame other than the parochial frame

Y **Alien measures**

Parameters in a moving (alien) frame as <u>calculated</u> from a stationary frame

Y **Alien speed of light S_A**

Speed of light as calculated (or assumed) from an alien frame

Amorphous

A substance with no shape or form

Y **Amorphous space D_ϕ**

A non-dimensional empty space

Y **Anglewarp**

Adjusted warp field caused by rotation and causing angular momentum

Y **Anglewarp equivalent**

The rotational equivalence principle

Angular momentum

Tendency of matter to continue rotating with nonzero angular speed

Angular reaction

The equivalent and opposite rotational response to any motion

Y **A-sea´s**

The pronunciation of "asies"

Y **Asies**

The plural of "asys"

Y **A-spec**

Abbreviation for "advanced speculation"

Assumption contrivance

An artificial (unnatural) assumption

Assumption creep

Stealthy increase in assumptions

Y **Asys**

An axiomatic system

Axiomatic systems

A logical system based on assumptions called axioms

Y **Base-line acceleration**

Minimum acceleration of m·a to accelerate slow moving objects

Y **Baseline force**

The initial assumption that F = m · a

Bernoulli's principle

An increase in speed results in decreased water pressure

Big Bang

Colloquial term for the creation event "CE" (beginning) of the universe

Y **Bilarities**

Motrons that have been mattified to gain mass

Y **Bilarity epoch**

The short period of matterfacation during inflation

Bootstrap

A solution to a circular demand

Y **CE**

Abbreviation for the "creation event"

Celestial solar equator

Projection of sun's equator onto the heavens

Centripetal force

The force needed to keep a rotating object in its orbit

Challenger disaster

Disintegration of NASA's Challenger spacecraft shortly after launch

Classical relativity

The current belief that all motion is relative (and can never be absolute)

Clumpability

Ability of galaxies to maintain their structure while the universe expands

Y **Cog**

Proposed distance unit for electromagnetic radiation (per second)

Constant c

Speed of light, previously a <u>universal</u> constant) for special relativity

Constant frame

The parochial reference frame, in which c is constant

Y **Contours**

Co-temporal curves

Y **Contra-warp**

Transfer of gravitational warp pressure from repulsion to attraction

Converse confusion

Conflation of an implication with its converse

Cosmic acceleration

The current expansion of the space network

Cosmic flatness

The Euclidean geometry of the space network

Y **Cosmic onion**

A 4-D model of space-time.

Cosmological principle

The belief that the universe is homogeneous and isotropic

Y **Creation event**

Modern name for the Big Bang

Dark Energy

Physical particles (fabrons) of the space network (fabric of space)

Dark Matter

Physical particles (warpons) of all warp fields (probably q-pons)

Deductive reasoning

Using logic with assumptions to prove conclusions

Density

The ratio of mass per unit volume

Density averaging

The use of a mathematical average for density approximation

Y **Density circles**

Using distance between concentric circles to represent angular speed

Y **D-flatness**

Dynamic (Euclidean) flatness (with respect to time)

Disturbing its symmetry

Loss of warp symmetry caused by motion

Y **Dole**

"Sequential fabron subdivision" analogy to governmental handouts

Y **D-pushing**

The ability of a warp field to maintain its density "D" while expanding

Drafting

Fuel advantage gained by one vehicle within the slipstream of another

Y **Dynamically flat (D-flat)**

Euclidean flatness that is maintained by continuous expansion

Y **Earth block**

Primordial energy structure anticipating role as future provider of matter

Y **Effective force**

The true force (for an acceleration) when adjusted for ellipsoidal afterwarp

Y **Ellipsoidal afterwarp**

The oval shape of the afterwarp causing "equivalent mass" adjustent

Epicycles

Complex circles of Ptolemaic astronomy (comparable to special relativity)

Epoch of inflation

Extreme expansion of the universe soon after the CE.

Equivalence Principles for gravitation, precession and nielucion

Similarity between warp forces and corresponding manual forces

Equivalent mass

Supposed "mass m adjustment" (at speed v) to compute $E = mv^2$

Y **Expanding for a reason**

"Space network expansion" controlled by D-pushing

Y **Expansion deficit**

Expansion rate required for flatness, minus universal expansion rate

Y **Fabrons**

The particle singularities that provide the fabric of space

Fermions

Elementary particles associated with matter

Y **First moment of stability**

The conclusion of matterfacation

Flatness

Characterization of the curvature of Euclidean geometry

Flatness problem

1st conceptual problem with "standard Big Bang model" of creation

Y **Force causes motion**

Force causes acceleration instantly; over time, it builds momentum

Y **Force indices**

Use of Lorentz transformation factors as force-effectiveness factors

Y **Forced relativity**

Special relativity/general relative combined (assuming universal c)

Y **Force-effectiveness**

Compensation for elliptical nature of afterwarp

Y **Force-factor**

Use of Lorentz transformation factors as force indices

Y **Fractalian**

Resembling or having the characteristics of a fractal

Frame of reference

A physical coordinate system

Y **Galactic contraction**

The effect of gravitation on universal expansion

Y **Galaxy pooling**

The propensity of galaxies to optimize their gravitation to survive

Gauge Boson

A massless particle that carries one of the fundamental forces

Y **Globacon**

The property of global consistency or invariance with time

Y **Globaconic**

Any structure or system with global consistency over time

Gravitation

Interaction between the warp fields of massive objects

Gravity

Attractive force between two massive objects

Gravity assist

Procedure for using gravity to increase speed in space maneuver.

Y **Greater Milky Way**

The massive but less visible outer portion of the Milky Way galaxy

Y **Happy tap**

A hypothesis unexpectedly explaining a known physics conundrum

Y **Heat pool**

The heat singularity that accumulated at O_0

Y **Heaven's gate**

The threshold at the beginning of time

Heisenberg Uncertainty Principle

Impossibility of knowing both the time and location of a moving particle

Horizon problem

2nd problem (after flatness) with the "standard Big Bang model" of creation

How it works

The narrative goal for understanding physical phenomenon

HUP

Heisenberg Uncertainty Principle

Hypercube

An octahedroid or tesseract

Hypergraph

Graph with one independent time variable and 3 space coordinates

Hyperspace

Space of more than three dimensions

Y **Hyperspective**

The perspective ln hyperspace

Y **Hypertate**

To rotate in hyperspace

Impulse

J = force times time

Induce

To infer a conclusion based on particular assumptions

Inertia

Propensity of motion in Newton's 1st law

Y **Internal warp**

An objects own warp field

Inverse square law

A property <u>diminishes</u> proportionally to <u>square</u> of the distance traveled

Isothermal expansion

Expansion of universe that does not cause temperature change

Y **Isowarps**

Afterwarp represented by "concentric" elliptical warp contours

Y* **Light speed**

The constant parochial* speed of light c

Y **Linear bent**

Cause of linear motion with independent anglewarp and orbitwarp

Linear momentum

See momentum

Y **Linear precession**

The linear equivalent (caused by gravity) of (angular) precession

Y **Linear reaction**

The 3rd law of linear motion (reaction)

Y **Linear straightening**

The linear result of cosmic photon pressure

Y **Maintenance reduction**

Minimal (Maintenance only) singularity reduction

Y **Manual-warp force**

Force such as precession with an artificial "distortion force"

Y **Margin inflation**

Minimal (maintenance) inflation

Mass

Proportionality constant between effective force and acceleration

Y **Mass image**

A visual representation of the massive extent of any massive object

Y **Mass/energy**

A property of existence that can be interpreted as mass or energy

Massive object

An object with measurable mass

Y **Matched curves**

Two or more simultaneous curves

Y **Matched treks**

Two or more simultaneous treks

Matterfacation

Cosmic conversion of energy to matter under extreme conditions

Measurements

Physical determinations by physically measuring

Measures

Mental determinations or calculations

Y **Medium of passage**

The space network (fabrons) that provide transport through space

Y **Micro-verse**

The tiny universe immediately following the CE

Y **Milky Way Proper**

The original "Milky Way"

Y **Milky Way Suburbs**

The dark-matter orbiting the Milky Way Proper due to nielucion

Y **MOAS**

The a**bbreviation for the "Mother of All Singularities"**

Model

Matching physical patterns with mathematical formulas

Momentum (Linear)

Property of afterwarp maintaining speed & linear bent

Y **Monopon warp ray**

Warp field with one internal warpon per q-fab, (e.g. photon beam)

Y **Motion Conjecture**

The five basic assumptions in natural relativity

Y **Motivated expansion**

Universal expansion due to D-pushing

Y **Motron density**

The distribution of energy with respect to the volume of space

Y **Motron Stability Conjecture**

Fixed fabron density regardless of cosmic expansion

Y **Motrons**

Warpons and fabrons collectively

Y **M-rule**

Massive matter maintains its frame

Y **Narrative Physics**

Understanding, beyond mathematical models, of particular phenomenon

Y **Natural behavior**

"Well understood" phenomenons are to be considered with suspicion

Y **Natural relativity**

A "simplification up-date" of special and general relativity

Y **Network bent**

Tendency of space to maintain constant density while expanding

Y **Network expansion**

The cause of universal expansion

Newton's laws

Three general properties of motion established by Sir Isaac Newton

Y **Nielucion**

Interaction between object's spinwarp and satellite's anglewarp

Y **Now universe**

The entire universe that is visible at "this" time

Y **Null-pons**

A single (quantum) fabron

Occam's razor

Notion that "better" theorems have fewer/simpler assumptions.

Y **Omega-max** or **ω-max**

Maximum angular speed corresponding to linear speed of c

Y **Onion scale**

The distance assigned to the common thickness of all cosmic onion peels

Y **Optional tagging**

New names given on the basis of newly recognized classifications

Y **Orbit advance**

Accelerated orbital progress caused by nielucion

Y **Orbitar momentum**

Tendency of matter to continue orbiting with nonzero orbitar speed

Y **Orbitar nielucion**

Newly discovered force that induces precession of orbits

Y **Orbitar warp**

The effect of pressurewarp on a nearby anglewarp

Y **Orbitwarp**

An adjusted warp field, caused by revolution, and maintaining nielucion

Origin

A basic coordinate system starting point

Y **Origin-specific system**

Coordinate system with a specified origin such as O_0

Y **Overwarp**

An external warp provided (by a massive object) to its superfield

Y **Parochial constancy c**

Constant speed of any and all local frame light speed

Y **Parochial light speed**

Speed of any and all local frame light speed

Particle Physics

The study of subatomic particles

Y **Photon pressure**

Cause of uniform distribution of fabrons throughout the universe

Planck length ℓ_P

The theoretically shortest measurable length

Y **Plane flattening**

Causing 2-D geometry to become Euclidean

Polarization

Alignment of the electromagnetic fields of photons

Pooling problem

Intense radiation preventing coalescing of normal matter

Precession

Interaction between one spinning object and the warp field of another.

Y **Pressurewarp**

Distortion of a warp field caused by the presence of a second warp field

Y **Principium axioma**

Axiomatic Principle: "Theorems of one system don't apply to others"

Puddling dispersal

Dispersing same content and volume into fewer locations

Y **Q-fabs**

A fabron singleton

Y **Q-mass**

A quantum (minimum quantity) of mass

Y **Q-pon**

Similar to a boson except with recognized Q-mass

Quantum energy

The minimum quantity of energy, energy associated with Q-mass

Quantum gravitation

A quantum (minimum quantity) of gravitation

Radial arm

Distance from axis of rotation perpendicular to arc of travel

Y **Radius of confidence**

Margin of error for distance in the space network

Y **Realm of regality**

Areas of knowledge that are given a pass on understanding

Y **Regal science**

Concepts that are treated as fiat

Relativistic velocities

Speed high enough to alter alien measures of c significantly

Relic radiation

The remnant of the energy from the CE (Big Bang)

Rest mass

The mass of a stationary object

Y **Revving**

Displacement of a warp field causing afterwarp, anglewarp, or orbitwarp

Riemannian geometry

A closed universe with negative constant that would eventually collapse

Right-hand rule

Using the right hand in a mnemonic to remember algorithms

ROP

Rate of progress (as distance divided by time of travel in any units)

Y **Rotational gravitation**

Possible name substitute for "precession"

Y **Ruffles**

Any disturbances caused by motion of a warp field

Y **Sequential matterfacation**

Matterfacation that occurs sequentially as from a "heat pool"

Y **Simplest Occam**

The narrative option with the fewest and simplest assumptions

Y **Simulated Afterwarp**

Artificial increase in an object's afterwarp

Y **Simulated Revving**

Gravitational ruffling-without-revving of like-directed warp on the far side

Singularity

An energy field of zero dimensions

Y **Singularity dole**

Release of individual fabrons from a fabron singularity

Y **Singularity reduction**

Reduction of a singularity of fabrons into individual q-fabs

Slingshot effect

Like "drafting" but using gravitation in space, (see linear precession)

Slipstream

The draft caused by an object moving through air

Y **SNA**

Abbreviation for space network acceleration

Space continuum

Three dimensional space

Y **Space energy**

Energy of the space network

Y **Space gaps**

Tiny spaces between the "strands" of the space network

Space warp

The version of warp fields imagined in special relativity

Y **Spaces network**

Fabrons of the universe

Space-time continuum

Math version of 4D space with time as the 4th dimension

Y **Span**

This refers to the distance between motrons

Y **Span average**

The central tendency for the lengths of a collection of spans

Special relativity

Version of relativity theory requiring uniform motion (zero acceleration)

Y **Speed angle**

The angle of the anglewarp that increases with the speed

Speed of light

An unmodified expression to include alien speed S_{A}.

Y **Speed slant**

Sufficient road slant to maintain a turn radius at a specific speed

Y **Spin revving**

Result of rotating an object to cause angular momentum

Y **Spinwarp**

Warp distortion of orbiting frame by the spin of the central object

Y **Spoke**

Average parochial span per second

Y **Stability**

The ability of fabrons to maintain fixed density while expanding

Stable

A property that remains constant over time

Y **Steadiness**

Ability to remain stationary

Y **Strongest heart**

Logical system with strongest context and narrative simplicity

String theory

Theory of physics dealing with the dynamics within hyperspace

Y **Subfield**

The warp field that is physically attached to an object

Y **Superfield**

The maximum contiguous warp field

Y **Temporal expansion**

Using time to maintain flatness

Tesseract

A formal name for the hypercube

The God particle

A hypothetical particle that assigns mass to its chosen objects

Top Quark

One version of the quark elementary particle as personified in **MU**

Trajectory

The path followed by a moving object

Y **Transflat model**

Original 2-D model of **MU** that models the dynamic flatness of D3

Y **Treks**

The trajectory of a photon at the exact time it passed through

Y **Turn-center coordination**

Adjusting multiple turn centers to become concentric

Type Ia supernovae

The detonation of a carbon white dwarf

Y **Underwarp**

The internal warp of a warp field

Uniform

A property that remains constant over space

Y **Uni-Verse**

A single verse refrain about life

Universal c

Light speed that is currently assumed constant in special relativity

Y **Universal stability**

The space network is composed of motrons with fixed density

Y **Universal wasting**

The (hypothetical) loss of material density due to expansion

Y **Vacuum**

Partial vacuum w/fabrons

Vacuum catastrophe

The huge discrepancy between measured vacuum density and field theory

Vacuum energy

An earlier name for the fabron energy of the space network

Valid

An argument in which the premises support the conclusion

Y **VAMOS**

Abbreviation for "variable motron separation"

Vector

A representation for motion that has both distance and direction

Y **Vectorwarp**

A compound warp caused by two or more warp sources

Y **Visible expansion**

The measurable component of cosmic expansion

Y **Visible universe**

Entire look-back "now" universe.

Visualization

A tool used to communicate behavior in the universe

Y **Warp displacement**

Change in location due to motion

Y **Warp field**

A field or span of warpons

Y **Warp field density**

A symptom and not a driver of flatness

Y **Warp Forces, f_w**

The forces of gravitation, f_g, precession f_p, and nieluty f_n

Y **Warp rate**

A measure of the relative loss of symmetry of any warp field

Y **Warp units**

A unit used to measure and expresses the value of a warp rate

Y **Warp value**

Quantity of "equivalent mass" provided to an object by its speed

Y **Warp-loss**

Nearside force of a pressurewarp

Y **Warpon**

Motron of matter that remains integrated within it giving it mass

Well understood

Natural behavior that should be considered with suspicion

Narratives

Name

Pages

About the Author

Photo by Keller N. Thornton

Professor Maxwell M. Hart has been a teacher and a full-time university professor of mathematics, physics, statistics and astronomy, collectively spanning over 50 years.

Among his earliest recognitions, from the Boy Scouts of America, he received the Eagle Scout Award, the God and Country Award and the Order of the Arrow award.

As an undergraduate student at Baylor University, he was recognized for his excellence in physics when he was inducted into the physics honor society (ΣΠΣ) in the fall of 1962. After attending three different colleges and changing majors twice, he finally graduated with degrees in education and mathematics with a minor in physics. He then received a graduate degree in mathematics. Before beginning his teaching career, he was aided by the highest possible ranking of 99 percentile on the National Teachers Exam. This ranking combined with his degrees in mathematics and physics allowed him a wide latitude in choosing his place of employment. He later received a postgraduate degree in computer science.

He has produced many papers and presentations over the years. His most recent publication in the field of physics, **MOTION UNIFICATION: THE NARRATIVE** is a result of several discoveries he made while still a college undergraduate before transferring to Baylor University. These revelations in physics and other discoveries in mathematics led him to many new breakthroughs. Some of the newly uncovered concepts seemed to be at odds with the physics and mathematics he was learning in college classrooms and also with what he was reading on his own.

Lacking confidence in his physics discoveries and other considerations caused him to concentrate his efforts in the study of mathematics and astronomy. After raising a family and regaining confidence in the nonconventional concepts he had developed, he began a concerted effort to finish his treatise on **MOTION UNIFICATION**. Due to the intense effort required in the completion of this physics manuscript, he is now finally able to publish it months after retiring from teaching.

Index

D

E

K

L

M

Q

R

S

T

Printed in the United States
By Bookmasters